RURAL SECOND HOMES IN EUROPE

Rural Second Homes in Europe

Examining housing supply and planning control

NICK GALLENT
The Bartlett School of Planning, University College London

MARK TEWDWR-JONES
Department of Land Economy, University of Aberdeen

Routledge
Taylor & Francis Group

LONDON AND NEW YORK

First published 2000 by Ashgate Publishing

Reissued 2018 by Routledge
2 Park Square, Milton Park, Abingdon, Oxon OX14 4RN
711 Third Avenue, New York, NY 10017, USA

Routledge is an imprint of the Taylor & Francis Group, an informa business

Publisher's Note
The publisher has gone to great lengths to ensure the quality of this reprint but points out that some imperfections in the original copies may be apparent.

Disclaimer
The publisher has made every effort to trace copyright holders and welcomes correspondence from those they have been unable to contact.

A Library of Congress record exists under LC control number: 99076649

ISBN 13: 978-1-138-70615-6 (hbk)
ISBN 13: 978-1-315-20197-9 (ebk)

Contents

Preface

This book developed from a project funded by Gwynedd County Council in 1996-97 that sought to identify the nature and scale of the second homes 'problem' in Europe. The work formed part of a European Union PACTE programme project on rural change experience within five regions in Europe that have additionally witnessed significant second homes growth. Our initial desk study on the second homes subject yielded a significant amount of literature covering a 30 year time span. Previous studies have tended to be devoted to an analysis of second homes either as an unwarranted aspect of rural lifestyles in specific regions, or else as a welcome contribution to a renaissance in the countryside with associated economic benefits. Wider problems experienced in the countryside form an extremely important context if policy planners and rural communities are to understand and attempt to ameliorate second homes as a problem issue. Rural areas have been subject to, amongst other things, massive socio-economic restructuring, a decline in agriculture and in rural services, and a lack of affordable housing for local people. These issues have underpinned and, to some extent, precipitated the second home phenomenon; all too often the wider problems being experienced in some rural areas have been obliterated by heated personal reactions against second homes. For some communities and political representatives, second homes have been viewed as the root cause of rural decline *per se*.

The significance of the second home problem within these wider rural issues is a subject that requires sensitive dissection. The negative aspects of second home growth have to be balanced against the positive aspects. The nature of rural problems more generally, including the socio-economic, cultural and linguistic issues indigenous to each separate community, need to be examined in assessments of second homes in particular areas. This, of course, makes the search for any European or national political 'solution' to second home regulation extremely difficult. We have written up the work from a British perspective but our desire to research second homes in other European states and include this material should make the final version of the book of relevance and interest to academics, policy analysts and planners across the EU.

The spatial scale of second home concern is a matter that has long intrigued us. We commenced our work as a response to one local planning authority's desire to consider the policy options available. Further research indicated that similar problems were being encountered in other European countries at the local or regional level. Then, while we were undertaking further research, the second homes issue received attention once again from national politicians in the UK. In the summer of 1998, two Private Member's Bills were introduced into the House of Commons seeking to regulate second home growth in socio-culturally and environmentally sensitive areas. Although both Bills failed to become law, the moves by politicians to debate the issue did generate an increasing interest from local planning authorities, members of the public, the countryside lobby, and the media, and we have been particularly encouraged by the number of letters and telephone calls received since this time supporting our work and requesting further information.

Within the context of the UK, we feel that the time is ripe to consider second homes within an evolving new political and institutional structure, following devolution, and discussions about the future form and role of the planning system. But we also hope the book will provide a useful contribution to more general on-going political and academic debate concerning the future of rural areas, housing demand and supply, leisure and tourism opportunities, and migration patterns across the European Union. Part I considers the central themes of the second homes issue, and addresses the origins of the debate, the scale and nature of the problem across Europe, the demand for second homes in rural areas, and economic, environmental and social impacts caused by second home growth. Part II looks at policy and practical responses within several countries of the European Union, and then considers what could be done through housing and planning mechanisms within a UK context. We conclude the book by offering some suggestions for further debate and research.

Acknowledgments

We are grateful to a number of professionals, colleagues and friends who have assisted us in our work over the last three years. We would particularly like to thank representatives of the five regions who participated in the EU PACTE Rural Change project for their information, data, time and responses to our constant queries: Henry Roberts and Iwan Jones of Gwynedd County Council; Chris Higgins of Highlands and Islands Enterprise; Liz Hawkins of Scottish Homes; Seppo Laitila and Martti Salminen of Savo Regional Council; Francisco Flavia Ramil of Galicia Regional Government; Eva Lopez Barrio of the University of Santiago; Pege Schelander and Yvonne Cervinus of BOSAM Association of Local Authorities; and Kjell Peterson of West Sweden.

Special thanks are also extended to Dylan Phillips of Cymeithas yr Iaith Gymraeg, Councillor Dafydd Iwan, and Elfyn Llwyd MP, for their assistance with issues concerning language, culture and community. We would also wish to acknowledge a number of colleagues at our employing institutions over the last three years for their friendship, good humour and support, at University of Manchester and University College London (Nick Gallent) and Cardiff University and University of Aberdeen (Mark Tewdwr-Jones). Gary Higgs, Nick Phelps, Sean White, Chris Allen, Rhys Jones, Philip Allmendinger, Adam Barker, Eleri Evans and Gary Hamilton provided excellent distractions at crucial stages of the work in taverns, coffee shops and curry houses but also assisted us academically in our work too.

We also thank Anne Keirby and her colleagues at Ashgate for their support over a long period of time and for being prepared to publish our work. Finally, we wish to record our appreciation to Michelle Roberts and Arlene Heron for their invaluable word processing skills.

Nick Gallent, London
Mark Tewdwr-Jones, Aberdeen

PART I

CENTRAL THEMES OF RURAL SECOND HOMES

1 Introduction

Introduction

Social change and restructuring at the national and regional level is the product of the complex interaction of a variety of underlying political, economic, environmental and social sub-forces. It follows, that an understanding of social change should be grounded in a comprehensive appraisal of its various components; an approach which necessarily transcends regional boundaries. This may be achieved in one of two ways: first, it might be desirable to identify broader patterns of change (viewed in terms of shifting employment rates, housing conditions or some alternative normative measurement) which may be attributed to component forces. Secondly, potential components might be initially identified with a view to developing a 'bottom-up' perspective on wider concerns. This particular study adopts the latter approach, examining in detail, the phenomenon of second home growth (and housing change) as both an indicator and element of social and more general *rural change* in the countryside.

For the opponents of this particular consumption good, second homes are often seen as a root cause of a range of social 'problems' (such as regional economic decline, rural depopulation or local housing shortages). Those who view second homes in a more favourable light tend to argue that their growth is simply symptomatic of broader social trends (reflecting modern working patterns, preferences in leisure consumption, improvements in personal mobility and higher wages in society alongside regional economic decline in importing areas); their presence in certain areas heralds change but does not underpin it. Between these polarised views is the notion that second homes represent at least a 'complicating' component in wider patterns of change, an issue clarified by Bollom (1978)

> It is clearly difficult to isolate second home ownership as a variable because of the other agents of social change which will be operating, but if we accept the view that, rather than being the cause or the symptom,

second home ownership is more an added complication of social and economic decline, then it will be an agent of social change (Bollom, 1978, p.121).

Those observers who have emphasised the detrimental cause-effect relationship between second home ownership and social decline have often failed to take full account of the indigenous social processes at work in the receiving region (that is, the region experiencing growth in the proportion of dwellings used as second homes); all too often, second homes are a convenient scape-goat for those local or national governments failing to promote successful economic growth policies. On the other hand, to view second homes only as a 'symptom' of change represents a superficial analysis based on the presumption that change of dwelling use or the development of new-built second homes has no local or national knock-on effects. Because of the relationship between second home issues and other related patterns of housing change (for example, in-migration and the changing housing choices of middle class groups or social exclusion) this is unlikely to be the case as the discussion in this book will show.

Aims of the Book

The findings in this book are based around a study of the 'second home' literature commissioned by Gwynedd County Council in July 1996. The study was supplemented with empirical research in the following two years and has four main aims. First, to consider the problems associated with the study of second homes (principally, defining concepts and acquiring accurate information) and to highlight the main concerns in terms of potential problems and possible benefits. Secondly, to examine the academic literature focusing on the growth and impacts of second home ownership in Britain and abroad. Thirdly, to highlight a European context and examine differing experiences and responses to second home expansion; and finally, to consider the need to regulate future growth and examine the best ways in which to plan for or control new development. Because this book is primarily based on a review of prior studies together with the advantage of some primary local research, our final aim will be to suggest preliminary recommendations which could be used to further analysis through field-work.

Past Literature and Themes

The English-language literature surrounding second homes flourished during the 1970s. There was a marked down-turn in academic output devoted to this issue during the early to mid 1980s but then a resurgence of interest in the 1990s which focused mainly on the 'overseas' element of second home ownership (see Hoggart and Buller, 1994a, 1994b and 1995; Barke, 1991). Abroad (particularly in the Scandinavian states), second home ownership was already being identified as a mature social phenomenon by the mid 1960s with the development of holiday villages to cater for that demand which could not be accommodated in the existing housing stock. In the study of second homes, the point at which the supply of 'surplus housing' (that is, vacant or dilapidated) is exhausted and second home seekers turn their attention to mainstream market housing or create a demand for new purpose-built development, may be viewed as the point of maturity. It is at this point that new demand may cause environmental concerns (because of a need to regulate the siting and standards of new development) or socio-economic concerns as seekers begin to compete directly against sections of the local population for existing housing. In Britain, this point appears to have been reached in the early 1970s (Downing and Dower, 1972, p.159) and explains much of the early concern surrounding second home growth (Bielckus *et al*, 1972; Pyne, 1973; Jacobs, 1973; Williams, 1974) and the more academic concerns at the latter end of the decade (Coppock, 1977; Bollom, 1978; Davies and O'Farrell, 1981). Recent work by Hoggart and Buller (focusing attention on British owners in rural communities in France) is not indicative of any saturation of the home market, but rather the widening property-price differential between Britain and rural France in the late 1980s (Hoggart and Buller, 1995, p.186).

Before a more extensive examination of the literature, it is worthwhile noting that there have been a number of 'traditional concerns'. In some studies, the entire range is given equal weighting whilst in others, particular issues are emphasised (such as geographical growth patterns or planning policy responses). In brief, the normal issues raised include: the problems of defining 'second homes' either for statutory planning purposes or accurate assessments of total numbers (most studies consider this issue); difficulties in collating accurate information including the types of potential information sources that might be used (including national population censuses, local planning studies or local government rating registers); mapping the spatial distribution of second home growth (see for instance,

Crouchley, 1976; Davies and O'Farrell, 1981); explaining and modelling spatial distributions of ownership (Crouchley, 1976; Barke and France, 1988; Barke, 1991); the costs and benefits for receiving regions which include negative and positive economic impacts (for instance South West Economic Planning Council, 1975), the environmental losses and gains through the conversion of existing buildings or greenfield development and socio-cultural repercussions which are often considered through the use of attitudinal research (Thompson, 1977; Bollom, 1978). Many of the more general studies end by examining the need for a policy response (either at the local or national level) towards the expansion of second home ownership; these responses often involve the encouragement of growth (to maximise benefits), the curbing of growth (to minimise costs) or a compromise solution which involves the regulation of supply and demand and the channelling of development. Increasingly, it is recognised that growth in the number of second homes is not an isolated phenomenon but connects to various other elements of social change and to the macro-economy. Therefore policy responses must be grounded in a full understanding of wider contextual concerns. These concerns are examined in a series of chapters which systematically examine the range of issues highlighted above.

Concepts and Data Sources

The basis of any social enquiry must be the recognition and definition of the core concepts. In the study of housing, for example, the core concepts might include the 'household' or the 'dwelling' (Kemeny, 1992). In the study of second homes (a subject which precariously straddles the boundaries between housing, planning and leisure), the core concept is the 'second home' itself. In some respects, the term is somewhat of a misnomer as academic studies tend to view the 'home' as an individual's principal node of social transaction and interaction and in this context a 'second home' is not a 'home' unless it is viewed as a point of secondary social transaction with a deeper symbolic meaning for its user. This is certainly the position adopted by Salletmaier (1993) who argues that 'recreation spaces' (that is, the secondary dwelling) are much more than just physical places and should be seen within the context of place identity (in much the same way as a first home), having a marked influence on the user's 'action potential'. In short, Salletmaier's conception of the second home is as a space (which is not physically limited in type) for recreation and communication that may be 'vital for the users' personal identities . Clearly, Salletmaier's definition borrows heavily from the work of Häbermas (1987, 1991) and is more concerned with the sociological justifications for demand than the search for a practical definition.

The debate surrounding the definition of 'second homes' (for planning and information collation purposes) was at its most intense during the 1970s and many of the difficulties experienced at that time are far from being resolved today (see OPCS, 1992). Coppock (1977) argued that an element of stereotyping and misconception has tended to confuse the definitional issue; for planning purposes, some local authorities consider static caravans as potential second homes whilst others do not. Similarly, ownership (that is, freehold tenure) is often implied as a criterion for definition but it may also be argued that a second home, like a main residence, may be rented or leased. On top of this, second homes may be used by more than one family but at what point should these dwellings be viewed as 'rented holiday accommodation'? The issue of definition is certainly no easier to resolve across the European Union. In Spain, for example, many owners of flats which are rented out claim that these are second homes in order to avoid tax on rental income; it is a lack of statutory guidelines which has permitted this situation to evolve (Barke and France, 1988, p.144). The first real attempts to confront the definitional problem occurred in the early 1970s and were often grounded in local experience and circumstances. Pyne (1973) arrived at a three-tier definition for vacation accommodation which might or might not be a second home:

1. *Second home*: a dwelling used by its owner and possibly other visitors for leisure or holiday purposes and which is not the usual or permanent place of residence for the owner;

2. *Holiday investment property*: a dwelling owned either locally or outside the county and not permanently occupied but let to holiday makers solely on a commercial basis;

3. *Club/institute/company holiday property*: similar to above but used only by club members or company employees and clients (Pyne, 1973, p.3).

Clearly, the principal criteria for defining a second home as such were *type* (recreational) and *frequency* (seasonal) of use. In the following year, Pardoe narrowed the definition further, stating that a second home was:

> ... a static property which is the alternative residence of a household, the principal domicile of which is situated elsewhere and which is used primarily by members of that household for their recreation and leisure

(Pardoe, 1974).

Pardoe's definition allowed the inclusion of static caravans but excluded touring caravans. Furthermore, the definition is more restrictive in terms of *who* may primarily use the dwelling for their recreation and *where* they have their first home. Second home owners are usually absent owners who reside some distance from the receiving area. The notion that they are, by definition, non-local and liable to import non-local ideas and values is just one of the stereotypes highlighted by Coppock. Elsewhere in Europe, there is no necessary presumption that owners are *distant* outsiders; in the Spanish Census of 1981 for example, a 'secondary residence' was simply a dwelling not used as the 'normal residence' but used seasonally or at specific times (Instituto Nacional de Estadistica, 1983, p.9).

More recent studies of second homes in Britain tend to follow (and often modify) the definition employed by Bielckus, *et al*, (1972):

> ... a property which is the occasional residence of a household that usually lives elsewhere and which is primarily used for recreation purposes (p. 9).

The Bielckus definition incorporates the concepts of distance (from first to second home), use type (recreational) and use frequency (occasional). In research for the Countryside Commission published in the following year, Downing and Dower followed the same definition (accepting the main criteria) but specifically excluded touring caravans, boats, whether houseboats or cruisers, properties on short tenancies and properties in major cities and industrial towns (Downing and Dower, 1973, p.1). Tuck (1973) added that the second home would be different from the normal domicile from which members of the household 'travel to work or school'. Some years later, Shucksmith (1983) chose to use the same definition but insisted that the 'occasional residence' should consist a 'permanent building' and thereby exclude static caravans (Shucksmith, 1983, p.174). This distinction between built second homes and static caravans is now clearly defined in much of the literature.

In defining second homes, many studies have emphasised the issue of tenure. The Dartington Amenity Research Trust's [Dart] study for the Countryside Commission for Scotland (1977) included a range of tenure types:

> ... a property owned, long-leased or rented on a yearly or longer basis as the occasional residence of a household that usually lives elsewhere (Dart, 1977, p.12).

The Dart team (including Downing and Dower) had first raised the question of tenure in 1973 in their report to the Countryside Commission south of the border but had failed to fully resolve the issue. In Scotland, it was decided that for statutory purposes, any workable definition would need to recognise that second homes could be occupied on the same tenure basis as main residences and that definition was fundamentally in the hands of the 'user'. They argued that irrespective of location, use type, frequency or tenure, a second home was 'a dwelling which is occupied by a household that also occupies another dwelling elsewhere, that other dwelling being regarded by the household as the primary dwelling' (Dart, 1977, p.78). The team pointed out that with such a definition, 'a head of household would be required to declare whether a dwelling was his first or second home on any occasion where the distinction mattered for statutory purposes' (Dart, 1977, p.5). This approach was also employed by Jacobs (1972) in his study of the growth of second homes in Denbighshire, Wales. Jacobs argued that the 'hidden character' of the phenomenon made it difficult and perhaps unrealistic to employ rigid definitions (Jacobs, 1972, p.4). Second homes, their type and distribution, are essentially a function of local circumstances and Jacobs argued that 'because the pattern of use might vary widely it was unwise to delineate a second home too narrowly [instead] it was left to the owner to decide whether the use of his property in Denbighshire fitted within these very broad guidelines [that is, a dwelling which was not the normal residence and was primarily used as a recreation space]'. Clearly, this type of sweeping definition allows for a broad interpretation of what constitutes a second home. However, the lack of a more robust definitional framework has caused difficulties both in Britain and elsewhere in Europe with the collation of data which is difficult to interpret and the problem of false registration.

Using the broadest definition (from Bielckus, *et al*, 1972) and a base-line figure established by Barr (1967), it was generally agreed that there were between 160,000 and 200,000 built second homes in England and Wales in 1978 alongside some 150,000 static caravans (Bollom, 1978, 1; Bielckus, *et al*, 1972, p.39). Bielckus argued that the figure for built second homes could rise to 750,000 by 1985 (this prediction spanned 13 years of change and expected growth). However, growth was transformed into market stagnation in the mid 1970s mainly because of economic recession and the impact of the Housing Act 1974 on the eligibility of second home owners for improvement grants. In fact, by 1985, it was noted in Hansard, that there were 221,000 built second homes in England

and Wales (Hansard, 1985). The point to be made here is that the definitional problems were far from being resolved by the time of the 1991 Census. In 1991, the total number of second residences in England and Wales was apparently calculated to be just 92,550 dwellings (OPCS, 1993, p.50). The definition employed on Census night ran as follows

> *[s]econd residences* were defined as company flats, holiday houses, weekend cottages, etc. in permanent buildings which were *known* to be the second residences of people who had a more permanent address elsewhere and which were unoccupied on Census night. This classification was applied even if the premises were occasionally let to others (OPCS, 1992, p.25).

The definition clearly excludes empty dwellings not *known* to be second residences and potential second homes enumerated as *occupied accommodation* on Census night. The margin for underestimating the total number of second homes is certainly enormous and patterns of incorrect enumeration may be further confused by instances of false registration; this was found to be the case in Spain (Barke and France, 1988; Barke, 1991). The most highly publicised case of false registration occurred in Austria in 1981 (Bennett, 1985). The Austrian Census is used as a benchmark for the allocation of funds to local government (Bennett, 1985, p.298) and those local authorities with relatively large numbers of second homes receive extra sums of money whilst authorities with large numbers of second home *owners* lose money. The underlying economic logic is that second homes are seen as a consumption good and an additional household amenity in the exporting region; in the importing region, on the other hand, they are seen as being indicative of economic decline and viewed as a deprivation indicator. The state supports importing regions at the expense of exporting regions. In 1981, many of those authorities with high numbers of owners encouraged these owners to register *in* their second homes on Census night (by offering incentives such as housing grants and loans), reducing the number of owners in the first home (exporting) areas and increasing their local government allocations whilst decreasing the allocations of the importing regions. The effects on local government were dramatic and the results of the 1981 Census, not surprisingly, were highly controversial. Bennett notes that 'false registration of people in second homes in the 1981 Austrian Census produced considerable losses of population from most of the main cities, especially Vienna' (Bennett, 1985, p.309). The gain in revenue to Vienna alone was in the region of US $2 million with significant sums of money going to other large cities. The net effect on the enumeration of second homes (and their observed distribution) is not noted by Bennett, who is primarily concerned with the macro-economic effect of false registration. However, it is clear that a lack of clear statutory guidance on what constitutes a second home leaves the system open to inaccuracy either because of abuse or honest mis-interpretation. Common to many national censuses is the problem of grouping together all occupied dwellings, even if they are not used as the main residence.

A clear statutory interpretation of what constitutes a second home will provide the platform for more accurate assessment and an essential starting point for developing effective control mechanisms (if these are deemed necessary). Any move towards new legislative regulation over second home ownership (such as 'change of use' planning control) will require a definition which is applied evenly. If local planning authorities are permitted to develop their own interpretations, this could result in future demand (and the associated costs and benefits of second homes) being shunted from one region to the next with no continuity in policy.

Information on Second Homes

The next major concern of many second home studies has been the lack o comprehensive and reliable data resulting in a 'plethora of ill-foundec opinion and conjecture' (Rogers, 1977, p.85). According to Dower, the need for a definition is matched by a need for systematic information on the subject (Dower, 1977, p.160). A shortage of information in the early 1970s was often attributed to the 'recent' nature of the growth phenomenon, the fact that information collected was often ancillary to a wider purpose (either housing or tourism), that second homes often form too small a sample (in housing and tourism statistics) for reliable analysis and the fact that second home owners may be averse to answering questions, particularly in the face of hostile local press (Downing and Dower, 1973, p.1). Despite these difficulties, there were a range of information sources drawn upon by the earlier studies. The 1971 Census for example recorded 'vacant' or 'owner absent' dwellings giving some indication of the potential number of second homes across Britain. In 1981, a 'second residence' category was included in the enumeration but this classification also covered student accommodation in private residences (OPCS, 1992, p.25); it is not clear whether the enumerators believed that students occupied these 'occasional' residences primarily for recreational purposes. However, despite the separation of these categories for the 1991 Census, it still seems likely that Census data substantially underestimates the number of second homes in Britain (see previous section). Perhaps a more significant source of information in the local studies were the Rating Registers which record the permanent addresses of second home owners for rating demands. These have been used substantially throughout second home studies (see for example Tuck, 1973; Pyne, 1972; Jacobs, 1972) and their use accounts for the particular concern with the origin of second home owners. It has been noted however that an extremely small number of owners attempted to make their second homes seem empty so as to escape rates demands. Other sources of information focusing on housing and population included electoral registers which, until the beginning of the 1970s, distinguished those people who owned property in the area but lived outside (and therefore, provided a way of cross-checking the Rating Registers). Records of local planning decisions and applications for improvement grants could also be used to define the use of particular dwellings. Downing and Dower (1973) also note that the Department of the Environment's 'family expenditure survey' has long provided an important source of data regarding patterns of household expenditure on second homes (Downing and Dower, 1973, p.3) although it takes a sample of just 350 households in Wales. The same authors noted that in Scotland, additional sources of information alluding to the ownership of second homes exist including summary housing statistics issued by the Scottish Office and local Valuation Rolls held by districts and centrally in Edinburgh. Of particular value in Scotland is the 'Register of Sasines', a comprehensive record of all property dealings required by Scots law (Dart, 1977, p.13).

Apart from data relating specifically to population, housing and planning, information on the growth of second homes in Britain may also be extracted from annual statistics pertaining to patterns of tourism. The 'British National Travel Survey' (BNTS) and the 'British Home Tourism Survey' (BHTS) were used by Downing and Dower in the both the study of England and Wales (1973) and the later Scottish study (1977). Again, in Scotland there are additional data sources including the 'Scottish Tourism and Recreation Study' (STARS) and local appraisals undertaken by the Scottish Tourist Board (Dart, 1977, p.18). It should be noted that any relevant data are ancillary to the broader purpose of gauging patterns of tourist activity (with particular emphasis on the regional economic impact). Because of this specific purpose, data on second homes collected as a component of tourism often differ substantially from those estimates derived from housing or planning statistics. Again, the definitional issue is the key problem. The Office for Official Publications of the European Communities classifies non-hotel, hostel, guest-house, camp-site, etc. accommodation as 'private tourist accommodation' sub-divided into 'rented rooms in houses', 'homes rented from private individuals', 'accommodation provided without charge by friends or relatives', 'own dwellings which comprise visitors' second homes' and 'other' (including tents on unofficial camp-sites or boats at unofficial moorings). Normally, only 'own dwellings' are treated as second homes but this label could perhaps apply equally to the middle three categories. In 1993, eight of the twelve member states failed to provide accommodation data which fit into these categories and the data which were provided appeared to underestimate the phenomenon; for instance, there appeared to be just 1,691 'second homes' in the whole of the Republic of Ireland (European Union, 1993, p.33). This problem with tourist information regarding second or 'holiday' homes is recognised in Europe. In 1995, the Official Journal of the European Communities carried an article admitting that the problem of classifying different types of holiday accommodation was preventing accurate and up-to-date data collection (Official Journal, 1995, c236, p.20).

The British studies of second homes throughout the 1970s tended to draw on the results of a variety of other national and local surveys. The market research organisation 'Audits of Great Britain Limited' conducted an annual cross-sectional audit of households based on a 35,000 household stratified random sample (Downing and Dower, 1973, p.3) in which the owners of 'other dwellings' of various classified types were identified. The results of a number of local studies also figured prominently in the literature at this time; perhaps the most cited studies were the Welsh studies of Tuck (1973) in Merionethshire, Pyne (1972) in Caernarvonshire and Jacobs (1972) in Denbighshire as well as other English-focused studies in the Lake District and Norfolk. These studies were usually sponsored, if not fully implemented, by the local planning departments and illustrated a growing concern for second homes as a complicating factor in local planning and housing issues. In Scotland, a number of local and regional studies were undertaken by the Highlands and Islands Development Board, the Crofters Commission and the Forestry Commission (Dart, 1977, p.18).

For the numerous individuals and organisations who examined second homes in the 1970s, it was apparent that the growth phenomenon had been a neglected subject partly due to the definitional problems and partly due to the difficulties experienced in acquiring accurate data (Williams, 1974, p.64). It was clear therefore, that a comprehensive research strategy, that would address these problems, needed to be developed by both national and local agencies. To date, this comprehensive strategy has not materialised and the problems of definition and data collection are as acute today as they were twenty-five years ago. The European Union has hinted at a need for a fresh approach to monitoring, particularly in the light of continued pressure from the Danish Government. The failure of individual governments to examine the changing situation in recent years underscores the fact that the issue of second home ownership has been neglected in mainstream political agendas, either because it is genuinely not viewed as a significant social phenomenon (perhaps because emphasis has been placed on urban rather than rural housing issues) or because accurate assessments of the current situation might lead to demand for immediate action.

Despite the problems associated with data collection, it was clear to observers throughout the 1970s that the growth of second home ownership and use was an increasingly important issue in Britain's countryside; the origin and nature of this growth is considered in the next Chapter.

2   The Origins of Growth

The Origins and Social Context of Growth

In Scandinavia and North America, the 'tradition' of owning a second
home (in the contemporary sense of the term) goes back to the 1930s. In
more recent years, 'increased personal mobility through higher rates of car
ownership, higher disposable real incomes and greater leisure time have all
led to increased demands for second homes' (Coppock, 1977, p.11)
throughout a number of capitalist and non-capitalist societies; these same
factors are emphasised by Downing and Dower (1973, p.23) who add that
Britain has been slower than its European neighbours to develop second
homes in the twentieth century because of urbanisation and the loss
through decay of much rural stock; economic depression also suppressed
demand until recently.   At the same time, the general growth in outdoor
recreation, the 'revolution' in environmental awareness and the growing
incidence of conflict between the owners of second homes and other groups
(particularly where the phenomenon has reached maturity and there is a
direct competition of interests) has raised the overall profile of the issue in
the public's eye. The growth of second home ownership is symptomatic of
social change across societies as a whole (affecting both importing and
exporting regions), which acts to generate the motivation for this particular
consumption good and dictates where effective demand may be realised. In
reality, growth in second home ownership may be viewed as just one of a
host of 'social invasions' which include urban gentrification and the
creation of commuter or dormitory villages in city-regions. In the broadest
terms, 'this great oscillation [the movement of urbanites into the country]
could almost be seen as the ebb tide of that great flow which brought (and
still brings) rural people into the industrial towns and cities over the last
two centuries' (Downing and Dower, 1973, p.30).
    Because the growth in second home ownership is in some way
symptomatic of social change (and once established, becomes a

complicating factor in subsequent social restructuring), an understanding of this change must go some way towards explaining the growth phenomenon. Socio-economic change is not spatially homogenous (Savage, 1989, 244) and Britain, like other nations, is a society characterised by locational inequality. It is this inequality between regions (or nations) which fuels the demand and growth in second home ownership. The notion of 'importing' and 'exporting' regions (that is, those regions which experience growth in the number of second homes and those which experience growth in the number of second home *owners*) was coined by Rogers (1977, p.93) and provides a useful framework for understanding the origins of the growth in second home demand. Across society as a whole, increases in personal mobility (through higher rates of private car ownership and the increased accessibility of many rural backwaters; Rogers, 1977, p.99), more disposable income and reductions in working hours (producing more leisure time) have been observed, particularly in the post-war period. These trends have been most apparent in the more affluent regions with economies increasingly based around new industries and financial services. They have been of lesser importance in those regions dependent on traditional and declining economic bases such as agriculture or extractive industry. Socio-economic change, therefore, has been expressed in the form of improving standards of living and population growth in the exporting regions and economic decline and depopulation in the importing regions. Variations in regional economic vitality obviously has an impact on house prices; in general terms, the affluent regions experience house price rises (which remain affordable) whilst the relatively poor regions experience falling house prices but with locals still spending a far greater proportion of their disposable incomes on housing costs than their regional neighbours. Economic decline brings diminishing employment opportunities and therefore generates pressure for depopulation. In Britain and elsewhere in Europe, this pattern of regional advantage and disadvantage often follows rural and urban cleavages (although not exclusively). Urban economic advantage makes second homes attainable whilst rural disadvantage (manifest in falling house prices, depopulation and a growing number of vacant or empty dwellings) provides an ideal focus for demand.

These socio-economic processes make growth in second home ownership possible but do not necessarily explain why growth has become a reality. The potential British demand for second homes in rural France in the late 1980s was outwardly the result of house price differentials, widened by the property boom (Hoggart and Buller, 1995). But this factor does not explain why *potential* demand was transformed into *effective* demand. Low rural house prices, rises in urban income and mobility have combined with a growing cult of nostalgia' for the countryside (Newby, 1980b) and the desire to escape the pressures of city life (Coppock, 1977, p.9). In order to understand the growth phenomenon, it is necessary to consider aspects of personal ['actor'] motivation alongside demand [or 'structural'] considerations.

Coppock (1977) has argued that personal motivations might include a desire to participate in some activity which requires access to rural resources, a wish to maintain links with a rural area where the owner or relatives originated, providing a place for holidays, securing an investment, conferring status or providing a place for retirement (Coppock, 1977, p.10). Wolfe (1977) adds that these motivations (producing leisure choices) are also shaped by the state education system which gives children in North America and Britain extended holidays during July and August (Wolfe, 1977, p.28). Second homes, unlike other recreational consumption goods may serve both short-term and long-term objectives; in the short-term, they provide an occasional recreational space and in the long-term, they may be transformed into a first home or a place to retire (Clout, 1977, p.57). This factor will certain influence the decision to purchase a second home. Research on those motivating factors which are responsible for the generation of effective demand has been carried out by Robertson (1977) who argues that purchase decisions are based primarily around 'anticipated utility' (relating to the functional value of acquisition as a recreational space or an investment). Once acquisition occurs, Robertson argues that further decisions centre around 'actual utility' (the realisation or non-realisation of expectations) and may involve the disposal of a dwelling which does not produce the expected gains (consequently, the tide of second home growth may expand further, stabilise or retreat). The third segment of 'utility evaluation' according to Robertson is 'projected utility' (Robertson, 1977, p.135). Following on from initial acquisition and the experience of ownership, the owner's knowledge is more complete and if the projected utility value of the second home is held to be high enough, some owners may anticipate retiring to their second home in future years (Robertson, 1977, p.136). This 'utility evaluation' model for explaining the motivating factors influencing changing patterns of ownership seemed to have some surface plausibility in Hoggart and Buller's study of British property owners in rural France. In that particular study, the authors noted that potential demand was generated by widening house price differential (between southern England and certain parts of rural France) in the late

1980s and that the motivation (or the 'anticipated utility') for acquiring French property was the perceived status attached to rural living and the 'rural dream' (Hoggart and Buller, 1995, p.180). However, on arrival in France, some owners found that the 'actual utility' of ownership was diminished by language barriers and a failure to integrate in rural communities despite the friendliness of the local population (Buller and Hoggart, 1994b, 208). Some of these owners subsequently decided to sell their French second home. This transition from anticipated to actual utility has been marked by a marginal decline in the number of British second home owners in France since 1991. Nevertheless, other owners with perhaps greater tenacity have managed to overcome linguistic and administrative barriers (these are the 'most fluent and established francophiles') and with greater 'projected utility', some expect to, or already have, made France their first home. Robertson notes that a 'better understanding of the decision-making process will increase the likelihood of developing successful models of patterns of second home development' (Robertson, 1977, p.136).

The patterns of social change generating the effective growth in second home ownership over the last fifty years have also influenced the types of socio-economic groups making acquisitions. The increasing possibility of lower and middle income households purchasing second homes has been viewed by some observers as a process of 'democratisation' involving the 'extension of privilege'. Clout (1977) notes that the predecessors of the modern second home were the country estates or the *châteaux* of English and French nobility in the 18th century. In post-revolution France, the *châteaux* were replaced by the weekend houses of rich provincials (Clout, 1977, p.47). In this century and closer to home, Pyne (1973) notes that although many studies indicate that second home owners 'follow managerial and professional occupations and consequently belong to higher income groups' (Pyne, 1973, p.12), since World War II, there has been a growing contingent of middle and lower class owners (Pyne was referring particularly to the old Welsh county of Caernarvonshire). In the Dart study of second homes in Scotland, it was demonstrated that second home ownership was spread widely across the social spectrum and although there was a particular emphasis on professional people, there was also 'considerable weight among non-manual and skilled manual workers' (Dart, 1977, p.33). As opportunities for ownership have filtered down the socio-economic hierarchy (as a result of economic growth, higher wages and increasing leisure time: see Robbins, 1930), second home ownership has ceased to be the exclusive preserve of the rich. Dower (1965a, 1965b) argues that the era of mass-ownership is part and parcel of the wider leisure revolution and likens this revolution to a 'fourth wave' breaking onto the countryside, the successor to the growth of industrial towns since 1800, the urban expansion driven by the railways and the more recent expansion of car-based suburbs. The 'democratisation' of second home ownership, a trend that finds its roots in the 1930s, may bring a new kind of equality to many urbanites, but it also brings difficulties. In Britain and elsewhere, this process is largely responsible for the recent 'maturity' of the second home phenomenon, where the supply of surplus rural dwellings is exhausted and second home seekers may be brought into direct conflict with local people for existing housing. This process was already well-advanced in many European countries by the mid-1970s (for example, in Sweden, Denmark, Finland and France) and it was recognised that 'given existing British trends [...British] second homes will reflect to some extent, the recent increases in social democratisation of second home ownership which is apparent abroad' (Williams, 1974, p.37). In a study of second homes on the Gower peninsula in Wales in 1974, it was shown that 44 per cent of owners had an annual income of under £6,000 (Williams, 1974, p.56).

Increases in demand since the war have gone hand-in-hand with a democratisation of second home use (Bielckus et al, 1972, p.143) and a continuation of this trend in the 1970s meant that the 'pool' of households possibly seeking second homes was potentially huge. Bielckus argued that patterns of social change were creating new opportunities (in all walks of life; in education, employment, standards of living and in leisure) and producing an 'extension of privilege'. However, he added that the extension of privilege for some might mean increasing disadvantage for others through the creation of new social tensions (associated perhaps with negative impacts of second homes in importing regions); as Coppock explained, 'one man's blessing is often another man's curse'. The extension of privilege may be associated with a widening privilege gap as equity between classes is not matched by equity between regions. Bielckus et al point out that:

> ... this concern for equity, particularly with respect to housing and building resources, is basic to any suggestions made... which may act as a guide for future second home development. It is especially relevant where existing premises are available for use either for first or second homes (Bielckus, et al, 1972, p.143).

Ireland (1987) in his study of the small Cornish fishing community of Sennen continues this theme further. By undertaking qualitative interviews he was able to add further social comment on the process of second home democratisation. Analysis of older residents suggested that pre-war and post-war visitors were very different in their socio-economic make-up (Ireland, 1987, pp.88-89) with the earlier visitors being of a 'better class' (they were clearly wealthy and travelled with nannies and chauffeurs). In the view of the residents, this 'better class of visitor' had greater 'respect for their host population' and because locals were in a subservient class position, 'physical and cultural boundaries were rarely transgressed' (Ireland, 1987, p.92). In contrast, those owners who arrived after the war were, and continue to be, more broadly distributed across the socio-economic spectrum and are 'more willing to cross cultural boundaries in their quest for property'. In effect, the pre-war visitors (from a time of rigid class structure and non-social democratisation) did not compete for the same property as locals (their second homes were purpose built, symbolising their class status) and did not influence the socio-cultural character of the receiving area (because social divides were not crossed). On the other hand, the post war group often compete for the same properties as local people (where surplus dwellings are absent) and do influence the socio-cultural character of the area. Social-democratisation brings the obvious result of mass-ownership and growth in demand. Less obvious is the inherent problem of greater economic and socio-cultural competition and friction between parallel social groups. The net result is that both the extent and the nature of growth are significant aspects of the second home phenomenon.

Clearly, patterns of socio-economic change in both the importing and exporting regions create the favourable supply and demand conditions described above whilst at the same time, combine with processes of personal decision-making, to generate motivation, turning potential demand into effective demand. The Dart team summarised this process in the following way:

> Demand for second homes... is not a thing that may be coldly calculated. It is not a simple factor of, say, income and accessibility; but rather a matter of basic impulse, turned by fashion and personal contact into desire and then into effective demand only if the right sort of supply is available (Dart, 1977, p.66).

The same patterns of socio-economic change have also opened up second home ownership to a broader spectrum of social groups with country retreats no longer being the preserve of the rich. However, whilst the 'extension of privilege' may herald greater economic prosperity for some, it can, as Ireland notes, also accentuate potential social problems in importing regions. The focus of the next Chapter is the type and distribution of second home acquisitions.

3   Ownership and Demand

Introduction

This Chapter covers two key areas of concern; first, it considers the geographical distribution of demand, examining the factors which determine the location of growth regions; second, it looks at the types of properties becoming second homes in the past and types of properties which may become second homes in the future.

The Distribution of Demand

Coppock (1977) argued that the spatial distribution of second homes in importing regions is controlled by a number of factors including the distance from major centres of population, the quality and character of the landscape in importing regions, and the presence of specific physical features such as the sea, rivers, lakes or mountains. Davies and O'Farrell (1981), for example, found that proximity to a beach was an important factor in determining the location of second homes in West Wales while Barke's study of Malaga Province in Spain (1991, p.16) also noted this tendency for second homes to concentrate on the coastal fringe. Other factors include the presence of other recreational resources, the availability of land (or property), the different climates of the importing and exporting regions (Coppock, 1977, p.6), and the availability of services (Barke, 1991, p.17). Because of the issue of accessibility, determined by transport infrastructure and car ownership, Coppock notes that second homes in England and Wales tend to be within 100 to 150 miles of major population centres although Gardavský (1977) notes that the building of highways in Czechoslovakia meant a greater and more varied spatial dispersal of second homes in the 1970s. The earlier interpretations of growth patterns tended to rely heavily on these factors to explain the pressures on particular regions. In addition, where the growth of *rural* second homes was observed, it was often noted that demand reflected patterns of economic decline and rural depopulation. In North Wales, for example, Bollom (1978) noted that the distribution of second homes (around Blaenau Ffestiniog and the Machno valley) was due to properties ceasing as first homes with the decline in slate quarrying. Areas of economic decline (as was explained in the previous discussion) attract an influx of second home buyers (Bollom, 1978, p.2) and if this in-migration and replacement of the existing population results in the inflation of house prices, further depopulation may occur (Bollom, 1978, p.2). For this reason, patterns of demand often reflect current patterns of second home ownership. This was true in France in the early 1990s, but not because of pressures on rural house prices. If properties, because of their type and location, are more suitable for recreation purposes than first home use, those second home owners selling on may sell to incoming second home seekers, generating an autonomous sub-market, superimposing new patterns of ownership on old (Hoggart and Buller, 1995, p.194). In addition to this particular process, 'colonisation' of second home owners may occur in some areas where existing owners inform friends of new properties coming on to the market (40 per cent of second homes were acquired in this way in Denbighshire in Wales; Jacobs, 1972, p.15). More generally, those factors affecting distribution highlighted by Coppock are seen to have particular relevance. Clout (1977) noted that in the early 1970s, many French second home owners were attracted to the Mediterranean coastline (Clout, 1977, p.53) whilst Cribier (1969) noted that 'second home hinterlands have been identified around French cities, with diameters increasing in relation to city size and consequent volume of demand, and chronologically in response to improvements in transport technology' (Cribier, 1969, p.55). Psychogios (1980) argued that geographically, 'concentric ring patterns' (of the type used in classical models of urban analysis) may describe the evolution of second homes acquired for leisure purposes (in terms of distance) but this simple pattern is complicated by a multitude of factors as tangible as resources or as unquantifiable as individual idiosyncrasies or tastes (Psychogios, 1980, p.38; Crouchley, 1976, p.3). The presence of 'water surfaces', both natural and man-made may also provide a focus for second home growth. In general Clout argued that the distribution of second homes must be seen in the context of:

> ... existing patterns of settlement, the volume of rural out-migration, availability of vacant housing, existence of social or cultural attractions for particular groups to particular areas, and the role of inheritance in the acquisition of second homes (Clout, 1977, p.55).

In France, *inheritance* is often a key factor in determining the acquisition and distribution of second homes; it is not rare for a city-dweller to inherit a rural home from grandparents (Clout, 1977, p.58) and this is likely to be true elsewhere in Europe where rapid urbanisation has occurred in the twentieth century. However, more contemporary studies in France (Buller and Hoggart, 1994a; 1994b) argue that more recent French generations have been increasingly cut off from rural origins and the process of inheritance is becoming less important. In the Balearic Islands, patterns of inheritance are still important but attitudes towards the land have changed and local people who retain ownership of their parent's or grandparent's land often move to main residences in the urban centres, turning their land into 'hobby farms' or weekend homes (Barke and France, 1988, p.144). In the Balearics and in Spain generally, a peculiarity in the second home phenomenon is the 'duality' in the distribution of ownership; foreigners tend to purchase properties on the coast in key tourist resorts (such as Palma and Manacor); that is, in areas with maximum leisure utility potential. Spanish nationals on the other hand are more likely to acquire smallholdings in the interior which provide weekend retreats. These distributions represent cultural objectives and preferences which often complicate patterns of second home ownership.

Of increasing importance in the distribution of second home growth regions is the role of advertising (Clout, 1977; Buller and Hoggart, 1994a) and the effect of planning control. Planning control will be examined in greater detail in later Chapters although it is worth noting here that planning control over second home development has tended to lag behind ownership trends in many European countries (Psyhogios, 1980, p.47). The factors outlined by Coppock and others can be used to explain the various rural or coastal regional biases of second home ownership. However, Rogers (1977) argues that national or broader regional trends should not obscure the fact that the distribution of second homes is also important at the local level 'for it is here that the contrasts and conflicts are particularly evident from the viewpoint of both landscape and social contact' (Rogers, 1977, p.89). At the local level in Wales, reports from Caernarvonshire (Pyne, 1973), Merionethshire (Tuck, 1973), Carmarthenshire (Hughes, 1973) and Denbighshire (Jacobs, 1972) suggested little evidence of any regular pattern in the location and distribution of second homes; they did, however, appear to show a preference among second home seekers for smaller settlements in these rural areas. However, in a study of Cemaes in West Wales, Davies and O'Farrell (1981) found no evidence of any reduction in the number of second homes with increasing settlement density although they did find that smaller settlements do tend to have a higher proportion of second homes. This could point to a lack of particular preference for small settlements on the part of second home seekers; however, it also highlights the potential for a greater second home impact in smaller communities (Davies and O'Farrell, 1981, pp.103-104). In second home use and ownership, local studies have tended to emphasise the importance of the motor car and its association with the types of holidays and recreational use that goes with second home ownership. Car ownership may have two impacts on the distribution of second homes. Locally, second home ownership is certainly linked to recreational opportunities and therefore the distribution of homes should be related to facilities; however, the possession of private transport militates against the need for close proximity to particular attractions or facilities (Pyne, 1973, p.20). Nationally, access to private transport should define the distance relationship between importing and exporting regions (and therefore between second and first homes). However, in the early 1970s at least, despite relatively high levels of car ownership, movement patterns from first to second home still tended to be highly local, even when observed at the national level (Rogers, 1977, p.90). In Denbighshire, for example, 62 per cent of second home owners lived within 40 miles of Ruthin (Jacobs, 1972, p.13). Downing and Dower noted that 'most owners have their main residence either in the same region as their second home or the adjoining region' (Downing and Dower, 1973, iv). Rogers noted that:

> Even with the limited amount of evidence available at present, it is clear that there are very definite and individual patterns of movement within the country, related to the dispersion of the urban population on the one hand and suitable areas for second homes on the other (Rogers, 1977, p.93).

The distribution of second home growth areas needs to be understood in the context of the interaction of the importing and exporting regions. In the first instance, a diversity in social processes create the conditions for supply and demand in each region. More visible perhaps is the fact that a population interchange (and the growth in second home ownership) may be encouraged by increased accessibility with the development of motorways tending to open up rural backwaters. The extension of the M5 and M4 motorways in South West England and Wales respectively, for instance, has meant increasing accessibility to the National Parks and Areas of Outstanding Natural Beauty (the same is true of the extension of the A55 Expressway in North Wales). Second home regions are not geographically isolated and the expansion of second home ownership is dependent on the importing region's social, economic and physical relationship with neighbouring 'feed' or exporting regions. Rogers argues that 'as with many other elements of the rural economy the mistake of viewing problems of second homes as though they were isolated from the rest of the region [or other regions] has meant that only part of the problem has been appreciated' (Rogers, 1977, p.99). However, whilst it is true that broader perspectives add much to the analysis of a particular phenomenon and help avoid disjointed and partially relevant conclusions, it may be worth noting the point made by Cloke (1985) that 'rural studies as a framework of study may be threatened if social science continues to espouse structuralist epistemologies with their aspatial connotations' (Cloke, 1985, p.2). The local view must be balanced against the national perspective. Clearly, the issues surrounding second homes demand this balance. Rogers notes that someone from Liverpool with a second home in Ruthin is in fact underusing resources in Liverpool during time spent in Wales (Rogers, 1977, p.100). Rogers adds, that 'in short, second homes should not be regarded as isolated rural phenomena with isolated rural problems, but as elements in the economy of the city-region which have substantial spatial and temporal variations' (Rogers, 1977, p.100).

The distribution of second homes therefore may be determined by a variety of social and physical factors which increase the attraction of the various importing regions. The location of exporting regions is also important and local second home areas must be seen in the context of the city-region framework; this accounts for the fact that many of the second home owners in rural North Wales originate from urban North West England (Crouchley, 1976, p.3). More generally, in 1972, between 20 and 30 per cent of all second homes in the UK were located in Wales with another core concentration in the Lake District with both areas drawing a large proportion of owners from Merseyside and the surrounding region (Bielckus, *et al*, 1972, p.101). In examining the distribution of second homes, it is also essential to bear in mind those personal motivations which encouraged demand in the first instance; these might include 'escapism', the need for 'relaxation' and the desire for a holiday location which fits in with personal taste (Psyhogios, 1980, p.23). Likewise, an issue of *New Society* (1967) highlighted Wibberley's assertion that 'too much education produces too many social isolates' (Wibberley, quoted in Barr, 1967) and with more educated people seeking second homes, patterns of distribution are increasingly grounded in personal motivations rather than in wider structures of socio-economic change. Their prediction and interpretation is becoming ever more complex.

Second Home Types

The second part of this discussion focuses attention on the types of dwellings used to accommodate second home demand. It has already been argued that shifting emphasis from surplus rural stock to existing mainstream housing and purpose-built second homes is indicative of a 'maturing' of the second home phenomenon and has been characterised by increasing concern over the impact of second homes. These particular concerns are examined in later Chapters.

One of the underlying causes of second home growth in importing regions is economic decline. With this decline, some dwellings cease being first homes and fall empty (Bollom, 1978, p.2); in the early 1970s, the Ministry of Agriculture estimated that the number of farmholdings in England and Wales had been declining at a rate of 10,000 units annually (Downing and Dower, 1973, p.23). These 'surplus' rural dwellings are the *initial* supply source for second home seekers, a phenomenon reported by Hoggart and Buller in rural France (Hoggart and Buller, 1994a, 1995) and by the Dart team in Scotland (Dart, 1977, p.61). Pyne (1973) notes that the early development of chalets was brought under control by the Town and Country Planning Act 1947 and the consequent demand for static caravans was curbed by the Caravan Sites and Control of Development Act. At this point, demand was transferred to inexpensive, empty and often derelict cottages (Pyne, 1973, p.1) left in the wake of rural depopulation. This demand was given added impetus by greater economic prosperity in exporting regions in the 1960s. In Wales, *vacant* farmhouses in agricultural areas and dilapidated terraced quarrier's cottages in slate-quarrying areas were the first types of dwellings to accommodate the demand for second homes at a time when the growth in ownership was seen as being purely symptomatic of social decline in the importing region. Coppock (1977) argued that second homes are normally of two types, either adaptations of pre-existing structures (that is, conversions) or new purpose-built dwellings (Coppock, 1977, p.8). Whilst this is broadly the case, the 'adaptations' can be further divided between those homes which were surplus to local housing need (often derelict) and other dwellings which may have been required by local families, but which (arguably) became second homes when locals were outbid in the property market. The latter dwelling may well represent a 'transitional type' between vacant properties and purpose-built second homes, which are likely to be converted to second home use in a period of non-market saturation before local government takes steps to divert demand from existing housing (by building new second homes) or seeks to control change of use (by acquiring and then letting, existing properties which come onto the market). Arguably then, the 'evolutionary' transition in dwelling types sought as second homes is from vacant dwellings, through existing mainstream housing, to purpose built units. However, this pattern is neither universally accepted nor applicable. Wolfe (1977) in his study of summer cottages in Ontario argued that profound changes had occurred in the types of dwellings used as second homes between the 1940s and the mid-1970s; in particular, he noted that 'the greatest change of all has been the growth in the number of second homes that are *not* purpose built. It is astonishing how many working farms have been abandoned throughout Ontario... and are now serving the inessential purpose of being a second home' (Wolfe, 1977, p.29). The development of second homes world-wide exhibits both 'contrasts and parallels' according to Bielckus (1977). In North America, for instance, the market has been commercially motivated whilst in Europe 'rural properties left vacant by migrants from agricultural areas have been taken over as second homes' (Bielckus, 1977, p.35). The dominance of early commercialism in North America accounts for large numbers of purpose-built second homes. In Britain and in Europe generally, developers have not, in the past, tended to become involved in second home developments and the more recent arrival of holiday villages and individual purpose built second homes is indicative of both growing commercial interest and a concern amongst planners to encourage tourist growth whilst diverting second home pressure from the existing mainstream housing stock. In 1972, Dower noted that:

> ... until now, built second homes were mostly concerted from existing buildings, without the planning authority having much say. From now on, they may be mainly new-built second homes, or newly sited static caravans or chalets, and the planning authorities may have a significant influence on what is built where, and how many are built (Dower, 1977, p.159).

In Scotland, the emergence of holiday villages (comprising chalets) in the 1960s was in part a result of commercial pressure and in part a response to planning concerns. Crofts (1977) noted that rising numbers of static caravans in rural Scotland was increasingly regarded as a planning problem in the 1960s (Crofts, 1977, p.103) and loopholes in the Caravan Act made their growth difficult to control. However, the regulation of built second homes and chalets was far more straightforward with a specified legal framework contained in the Town and Country Planning (Scotland) Acts and the Building Standards Regulations (Scotland) Act 1963 (amended 1967). In certain regions, and particularly in Argyll, it was decided that the 'blight' of caravans was such that the best way forward was to 'substitute' them with more permanent structures (which could be better regulated) and as a consequence of the 'substitution' process, holiday villages were created with the advantages of controlled design specifications and improved siting (Crofts, 1977, p.111).

The transition in second home types and the move towards mainstream housing and purpose-built dwellings (as sources of surplus housing have diminished) has generated much of the concern over the last 20 years for regulating second home growth through the use of planning control. By 1976, for example, the supply of vacant dwellings appeared to have all but evaporated in Scotland (Dart, 1977, p.6) and so revived demand (after the economic recession of the mid-1970s) would mean direct competition for existing dwellings or would involve new building and a structured response from local planning authorities. Despite these trends, the conversion of derelict or vacant housing into second homes may still be seen as beneficial, particularly if buildings are *preserved* by their change of use (Dower, 1972); studies by Hoggart and Buller in rural France have shown that second home owners have been the driving force behind significant improvements in the rural housing stock (Hoggart and Buller, 1995, p.196). In reviewing prior studies, it is often impossible to distinguish between the conversion of surplus and mainstream housing. In the South West Economic Planning Council's study of second homes in the South West of England (1975), it was found that a sixth of the region's second homes were purchased as new (many being purpose built) while the remainder were acquisitions of existing dwellings. Significantly, 68 per cent of these acquisitions had full amenities when purchased, clearly suggesting that they were not derelict properties; 30 per cent of owners made improvements and 11 per cent with the use of grants (although the rules governing improvement grants for second homes were amended in the Housing Act 1974 (Part VII, s60); see later discussion) (SWEPC, 1975, p.10). Pyne's study of second homes in Caernarvonshire (1973) showed that 74 per cent of second homes were unimproved or modernised existing dwellings (Pyne, 1973, p.10) and in Wales as a whole, most of the stock of built second homes had been created by change of use (Downing and Dower, 1973, p.8), although by the early 1970s, Wales had the highest concentration of static caravans across Britain, a fact that provided the focus of much disquiet (Llywelyn, 1976). The early studies demonstrated that by the early 1970s, the demand for second homes was increasingly focusing on the mainstream property market and the acquisition of derelict properties may have been declining whilst new-build remained marginal. For this reason, the 1970s onwards may be seen as a transitional phase in the types of properties being used as second homes, reflecting a fall-off in the supply of empty or derelict properties (Pyne, 1973, p.10) and the lesser importance of new build in accommodating demand at that time (although Williams points to a 'small number' of purpose built developments emerging across Wales at the beginning of the 1970s (Williams, 1974, p.12; see also Cymdeithas Yr Iaith Gymraeg, 1971). In response to this trend, Downing and Dower argued that:

> While second homes consist merely of properties no longer wanted as first homes, they cause relatively few problems, but when they grow beyond this, political, social, economic and environmental problems arise (Downing and Dower, 1973, p.32).

This phase in the second home phenomenon saw a growing concern that locals were having to compete directly with outsiders for mainstream housing (particularly for small dwellings with low rateable values which may have been suitable for first time buyers; Jacobs, 1972, p.9), the result being that inflationary pressure was being exerted on house prices and locals were being left without homes. These particular concerns are detailed in later Chapters. Another concern emerged from the Housing Act 1980. Because new purchase rights were awarded to local authority secure tenants (through the right-to-buy), it was believed that former council houses in the most attractive rural areas would become the next target for second home seekers, accentuating the rural housing shortage. As a result of this fear, the Government placed restrictions on the resale of council houses in certain designated rural areas in England and Wales (s19) and certain local authority pre-emption rights on resales in Scotland (Williams and Twine, 1994, p.193). In a study of the resale process in Scotland by Williams and Twine (1994), it was shown that of 8,249 dwellings sold to sitting tenants by 1991, 723 (or 9 per cent) had been sold-on. The study considered the extent to which resales became second homes; only four properties were confirmed as second homes whilst another seventeen may have been converted to this use. Clearly, most of these former council houses became the sole residences of their occupants with three-quarters of purchasers being local. The general conclusion was that affluent outsiders do not compete for these properties and:

> ... despite the fears raised concerning second homes when the right-to-buy legislation was drafted, the evidence from the case study areas clearly

indicates that resold council houses are not being used as second homes to any significant extent (Williams and Twine, 1994, p.207).

This study has particular resonance today with the introduction of purchase grants (in the Housing Act 1996) for the tenants of registered housing associations in England and Wales. Despite the exemption of dwelling in certain sizes of rural community, there must be some fear that housing association dwellings, lacking the regimented design standards of 1950s and 1960s council housing, will be an attractive proposition for second home seekers. An article *in The Independent* newspaper (17 August 1995) described how 12 cottages built in the Lake District Village of Rothwaite by Home Housing Association in 1953 had been sold to tenants in the 1980s under the right-to-buy (which originally affected association secure lettings). By 1995, 11 of the 12 cottages had been sold-on by their original tenants (with significant profit) and most are now second homes. Clearly, such anecdotes fuel fears that houses which should be serving the needs of local communities are falling prey to the expansion of the rural leisure industry.

In combination with patterns of overall second home growth and the social democratisation of ownership, it is the changing nature of the actual dwellings becoming second homes which has generated (and continues to generate) so much concern and social comment. In earlier phases of the growth phenomenon, the demand for second homes had little impact on the *effective* national housing stock and the conversion of derelict dwellings was often seen to bring local benefit. However, 'it is when the demand for built second homes exceeds the supply of housing which is genuinely surplus to local needs that strains appear' (Dart, 1977, p.62); at this point, demand for mainstream housing or new development are likely to create social and environmental tensions for a wide variety of reasons.

Modelling Future Growth

Before moving on in the next Chapter to consider the 'tensions' previously referred to, it is perhaps worthwhile considering those studies which have looked at how future trends in the second home growth phenomenon might be predicted through the use of modelling (Davies and O'Farrell, 1981; Crouchley, 1976). Most of the prior studies concerned with second homes agree that the growth in ownership experienced since the World War II is set to continue, particularly given the greater scope for development in

Britain compared to other European countries, which are still experiencing growth but where the second home phenomenon is certainly more advanced. Afterall, second home ownership is not just a passing phase in outdoor recreation (Williams, 1974, p.37).

Modelling and Prediction

Wilson (1974) has argued that planning should be fundamentally concerned with analysis, design and policy in that specific order. A concern for developing policy to deal with the second home growth phenomenon has created a need to analyse growth patterns in such a way as to provide a platform for effective and relevant policy decisions. Crouchley (1976) attempted to formulate a seventeen-variable model which could assist in such decision-making. He related the distribution of second homes in Denbighshire against agricultural intensity, the provision of public services, recreation facilities, the motivations of second home seekers, land ownership patterns, existing patterns of second home ownership, the decline in employment opportunities and rural depopulation and a number of other variables which he identified, from other studies, as potentially affecting distribution (Crouchley, 1976, p.12). Crouchley argued that the 'aggregated motives and factors which explain the spatial distribution of second homes can be grouped and transformed into a set of indices' (Crouchley, 1976, p.14) which can then be examined using multiple regression techniques and principal components analysis. From this type of data exploration, Crouchley hoped to arrive at a predictive model (Crouchley, 1976, p.60). In terms of the most significant indices, he concluded that patterns of future growth were likely to occur in areas with lower dwelling and population densities, lower accessibility indices and lower agricultural grades. That is, in parts of existing importing regions which are more isolated (although generally accessible) and where land is less likely to be in agricultural use. In Denbighshire, Wales, Jacobs noted that there were more second homes in areas of lower soil quality, rural depopulation and closer proximity to Merseyside (Jacobs, 1972, p.7). However, Crouchley's predictive model was limited by the percentage of variation in distribution patterns which were left *unexplained* (Crouchley, 1976, p.76) mainly because of the high level of data aggregation, the scale of the analysis, poor operational use of indices, errors in measurement, random variations and the failure to make the model temporally specific. In fact, the model was able to explain fewer variations than earlier models

32

developed by Aldskogius (1967) and Burby, Donelly and Wiess (1972). Crouchley concedes that models of all types tend to be better at predicting the behaviour of parts rather than wholes. The second home phenomenon, in contrast, is the product of socio-economic changes occurring at the national level, and not simply in the importing region. Clearly, this fact in itself has implications for the way in which policy to deal with growth should be developed. Should, for example, policy seek to sever the flow of second home seekers from exporting regions, adopting a prohibitive stance? Or should, on the other hand, a more holistic approach be developed which recognises the legitimacy of urban leisure demand and also the need to protect rural communities from the potentially negative impacts of growth?

The locational analysis attempted by Davies and O'Farrell (1981) was slightly more effective in producing particular policy recommendations. Their analysis of second homes was based upon cross-classification and regression analysis and found that growth patterns (around Cemaes in West Wales) had been determined by proximity to the nearest beach and a tendency to locate in smaller villages (Davies and O'Farrell, 1981, p.103). However, they also found that increasing distance from open land and higher settlement densities did not seem to deter potential second home owners from purchasing properties. Two key policy implications sprang from this analysis; firstly, second home seekers might be attracted to developments built at a comparatively high residential density (that is, in the form of holiday villages) and secondly, the location of demand suggested that housing policy could not deal separately with 'the housing needs of the local community and the pressure for second homes arising from outside the local area' (Davies and O'Farrell, 1981, p.107).

These types of studies suggest that under certain conditions, and where growth has developed in a particular way, second homes are likely to have social, economic and environmental impacts. It is easy, in this context, to point to the need to control growth, particularly through the use of the planning system which has proved relatively effective in controlling other types of development in the countryside since 1947. However, planning is not simply about control and restricting the expansion of particular activities. The encroachment of town into rural backwaters means that 'a stage has been reached when there is virtually no use of the countryside to which someone will not object' (Williams, 1974, p.6). But these objections should not obscure the fact that planning has two legitimate objectives. Certainly, the first of these is to minimise the potentially negative impacts of new development. However, the second objective, which demands equal emphasis, is to ensure that possible benefits inherent in new development are maximised. For this reason, the costs and benefits of all development types (including second homes) need to be considered before policy responses are formulated.

4 Economic Costs and Benefits

The Analysis of Costs and Benefits

The analysis of the various 'costs' and 'benefits' (usually affecting the *local* economy, society or environment) associated with the growth in second home ownership has been the linchpin of many of the prior second home studies. These studies have tended to recognise that comprehensive and quantitative cost-benefit analyses of the second home phenomenon are 'rendered meaningless' by the great number of necessary arbitrary assumptions (Pyne, 1973, p.21) relating to patterns of local expenditure or the impact on community life. For this reasons, the normal way in which to consider the impacts of second homes is to list the various factors, examining the relative importance of each in turn. It must be emphasised, however, that 'whether or not the outcome [of second home growth] is regarded as positive or negative, the impact for communities can depend heavily upon local circumstances' (Hoggart, *et al*, 1995, p.178).

House Price Inflation

Perhaps the main concern of those opposed to the growth in second home ownership is the supposed impact on local house prices where locals and 'newcomers' compete directly for mainstream housing stock. Bollom (1978) notes that in North Wales, 'younger people who wish to purchase properties may be frustrated in their attempts because of the greater purchasing power of potential second homers in the local housing market' (Bollom, 1978, p.112). However, the notion that second home seekers and locals compete for the same housing has not received universal endorsement and many observers contend that 'aspiring second home owners and local people compete for different forms of property' (Rogers, 1977, p.99). Whilst this is certainly the case in many areas (with demand concentrated in the non-effective housing stock), there are case studies

35

which reveal evidence of direct competition and upward pressure on house prices and these studies were becoming more prolific throughout the 1970s as the supply of surplus dwellings dwindled (Dart, 1977). In 1981, Shucksmith noted that:

> The essence of the housing problem in rural areas is that those who work there tend to receive low incomes, and are thus unable to compete with more affluent "adventitious" purchasers from elsewhere in a market where supply is restricted (Shucksmith, 1981, p.11).

Shucksmith argued that a free market in rural areas allocates to the most affluent whilst indigenous populations are left with the poorest and cheapest accommodation (if any). The crux of the problem is that insufficient houses are built and these go to affluent outsiders (for first or second home use). The real problem, according to Shucksmith, was that whilst rigid planning restrictions on land supply 'reinforce the inequitable consequences of a free-market allocation and lead to socially regressive distributional consequences', non-intervention has an equally negative impact and it has long been established that intervention is required to deal with efficiency and inequity problems (Shucksmith, 1981, p.13). Clearly, outside competition has a potential impact on property prices which may or may not be realised; outside competition however, may be viewed as only part of the problem with low rural wages largely contributing to the inability of locals to compete in a property market differentiated by more than just two consumption groups. The theory that the bidding up of house prices by second home owners contributes to rural depopulation is supported by the work of Bennett (1976) who examined this phenomenon in the Lake District (before the arrival of the Section 52 occupancy rulings). Pyne (1973) and Tuck (1973) have also acknowledged that rising house prices have been associated with the growth in second homes in some areas. Downing and Dower (1973) are more cautious, explaining that it is the combination of direct competition, economic depression and low rural wages that creates a housing affordability problem for rural households; they concede, however, that competition for housing in small towns and villages (as opposed to more isolated dwellings) can act to bid up prices (Downing and Dower, 1973, p.31). One of the more comprehensive studies of this relationship was carried out by Clark (1982) in a study, again of the Lake District, which compared different communities, house prices and patterns of second home ownership. Clark found robust evidence to support the argument that second homes do exert inflationary pressure on property values. The work of de Vane (1975), on the other hand, failed to reveal any *regional* empirical evidence to attribute local rises directly to second home ownership and it is clear that only local anecdotal studies have been able to demonstrate a lucid relationship (noticeably Jacobs, 1972). Jacobs tackled the issue head-on in Denbighshire in Wales; he identified areas where the sale prices of second homes (and surrounding housing) had soared in recent years (quoting from studies in the Peak District and the Lakes) and claimed that the same trend was discernible in North Wales. In Denbighshire, prices had risen by 9.3 per cent between 1962 and 1972 compared to a national rise of 6.7 per cent for existing properties over the same period. There was certainly local evidence that second home owners were competing against first time buyers for smaller properties, accounting, in part, for the differential between local and national house prices. (Jacobs, 1972, p.42). On Anglesey, a study by Jenkin (1985) of fifteen parishes with the highest proportions of second home ownership revealed that the 'chances of first time buyers securing a home in these parishes was bleak' (Jenkin, 1985, p.41) with only 1 in 25 houses valued at below £16,000 compared to 1 in 10 in the remaining parishes on the island. He concluded from this that claims put forward by The Welsh Language Society (1971), that is, that houses go to the rich with a subsequent inflation of surrounding property prices, are well-founded. What Jenkin failed to take into account was that those factors attracting second home owners to particular parishes may also have drawn more affluent locals. It is wrong to suggest that all local house price variations can be attributed to second home ownership in the same way as is wrong to assume that all 'locals' survive on low incomes. Shucksmith (1990) has noted that competition for housing occurs between a range of income and social groups and that crude local-newcomer dichotomies should perhaps give way to more nuanced typologies of competitive consumption groups developed by Saunders (1984), Ambrose (1974), Pahl (1966) or Dunn, Rawson and Rogers (1981). Jacobs (1972) is more sensitive to the fact that different groups compete for different housing. He notes, for instance, that increases in the cost of an isolated cottage has little impact on local families seeking housing as these dwellings are often not ideal places in which to bring up a young family. Similarly, it may be the case that when properties are on the market, altruistic locals may give first choice to friends and relatives. However, these offers are often not taken up by young people who gravitate to larger key settlements due to the greater availability of services and employment. A profile, undertaken by Jacobs, of the Llanrwst employment exchange between 1952 and 1969 revealed that scattered employment (i.e., jobs on isolated farms and in small villages) had declined far more than absolute employment, a trend that is likely to be repeated in numerous rural areas. For this reason, 'workers follow the jobs and migrate to the new centres of employment' (Jacobs, 1972, p.44) which may mean that they do not always compete with second home seekers for isolated properties or properties in small villages. Examples of direct competition between local households and newcomers seeking second homes certainly occur but may only be identified by fieldwork in particular locations. Recent research by Cloke *et al* (1994) in twelve case-study areas in rural England concluded that:

> There did seem to be widespread occurrences of affluent in-migrants outbidding local people for the limited available housing stock (Cloke, *et al*, 1994, p.161).

Although this phenomenon is not universal, it might be argued that it is becoming more commonplace as the supply of isolated surplus dwellings is depleted. But, again, this pattern will not be repeated everywhere as different localities are characterised by contrasting housing stock mixes and different patterns of population change (Cloke, *et al*, 1994, p.161). Cloke *et al* add that an 'established culture of uneven competition' if often the hall mark of countrysides subject to restrictive planning; housing supply rather than outside competition is the principal component in house-price inflation. However, this fact does not deter some observers from claiming that all increases in rural house prices (particularly in Wales) and the number of homelessness cases accepted by local authorities are directly attributable to second home purchases and the influx of permanent English residents (Pilkington, 1990, p.19).

Initial Acquisition

In comparison to the issue of house price inflation and direct competition between locals and second home seekers, other economic impacts appear far less controversial and perhaps even bland. However, each of the additional impacts is contentious. Many of the prior studies have considered the local effect of the initial acquisition of second homes (Dart, 1977; SWEPC, 1975; Pyne, 1973; Shucksmith, 1983; Jenkin, 1985). Jenkin (1985), for example, notes that that the acquisition of properties by outsiders may generate income in the local economy, with work for solicitors, estate agents and surveyors along with potential profit for the vendor.   However, these economic benefits are dependent on money accrued being reinvested locally (producing a multiplier effect; Jenkin, 1985, p.14). On this issue, the study by the SWEPC (1975) argued that in terms of local income 'expenditure on the purchase of second homes will usually be insignificant' (SWEPC, 1975, p.18) as profits will tend to accrue to just one person or household (the professionals employed, particularly solicitors may be hired in the exporting region). In the worst case scenario, it is even possible that the vendor lives outside the area and the property changing hands was already a second home (these autonomous sub-markets will produce little economic benefit through acquisition) (Pyne, 1973, p.21). In 1972, the SWEPC estimated that £1.6 million was spent each year on purchasing second homes in the South West of England and half of this total was spent in Cornwall (SWEPC, 1975, p.18). Initial acquisition may have some economic benefits in particular localities, especially if second home purchasers generate fresh demand in parts of the property market and stimulate new construction (Pyne, 1973, p.21). In contrast, acquisition may have an adverse effect on the property market, reducing the availability of rented housing, as local owners decide that selling rather than renting out property is a sounder financial proposition. The reduction in rented rural accommodation, however, certainly owes more to the effect of successive Rent Act legislation since the 1920s and the more recent introduction of the right-to-buy than the arrival of second homes (Shucksmith, 1983, p.179). In France, Hoggart and Buller (1995) have noted that many vendors were unable to sell their properties before the British buyers arrived; acquisition on the part of Britons and subsequent profits for the French vendors meant that local people could afford to move into new homes or make improvements in their existing dwellings (Hoggart and Buller, 1995, p.189). The net effect of initial acquisition is difficult to judge and quantify but it is clear that benefits or costs in different regions will depend on the types of dwellings purchased as second homes. If mainstream housing is acquired, the benefits for the local vendor may not be shared by other members of the local community.

Property Speculation

The issue of second home property speculation has received far less attention in the literature than other economic aspects of second home growth. Property speculation in the second home market was a growing problem in the early 1970s and was linked to the availability of improvement grants. Business-minded people were able to buy derelict properties which they could renovate using the improvement grants administered by local government. These properties could then be sold-on with substantial profits and the cycle could then be repeated. Clearly, in this situation, there is no benefit for local people unless they are directly involved in the speculation process. However, this practice was brought to an abrupt end in 1974 when the Housing Act introduced measures to make second homes ineligible for improvement grants (Pyne, 1973, p.23); this move pre-empted calls from some local authorities to reinstate the resale time limit on properties renovated using local authority grants (Pyne, 1973, p.44). Although second home speculation has not been a particular problem in Britain, it has caused a certain amount of disquiet abroad. In Greece, for example, a Royal Decree in 1967 allowed *Lyomeno* vacation houses to be erected anywhere on land where no 'Master' plan existed. This move was supposed to open up the chance of owning a second home to middle and low income families, accelerating the process of social democratisation in ownership. What it achieved instead was a huge rush in land speculation led by large firms which subsequently erected more expensive vacation accommodation for households in the upper income brackets (Psychogios, 1980, p.95).

General Housing Stock Improvements

The general improvement of properties purchased as second homes supposedly has both a general economic impact and leads to the improvement of the overall quality of the rural housing stock (Hoggart and Buller, 1994a, 1995). Buyers of second homes have tended to be far less concerned with the condition of properties than those looking for main residences; this has made the issue of improvement particularly important. Obviously this has much to do with the initial purchase price and the availability of surplus properties. However, both Jacobs (1972) and Hoggart and Buller (1995) have shown that many second home buyers have a market preference for older properties which they can improve and apply their own tastes and ideas to (Jacobs, 1972, p.10). This point is demonstrated by Jacobs. In 1970, the Forestry Commission put 14 surplus properties on the market in Denbighshire, Wales. They received 4,000 replies (using a closed bid system) and 13 of the properties were bought immediately; the only one not selling straight away was *not* in need of repair. Before 1974, the benefits accruing from the improvement of properties was confused by the use (often considerable) of local authority improvement grants on the part of second home owners. Jacobs (1972) showed that second home owners represented 6 per cent of the property owners in Denbighshire but claimed 10 per cent of the grant expenditure. He conceded that this direct comparison was not necessarily fair as many of these second homes had some of the lowest rateable values in the county and were in serious need of repair. However, Jacobs pointed out that 'second home owners are 1.7 times more likely to apply for improvement grants than people who are permanently resident in the county' (Jacobs, 1972, p.23). Pyne (1973) showed that 70 per cent of second homes purchased in Caernarvonshire were old and in need of repair; when grants were available, the cost of these repairs was partially incurred by local government, but it was estimated that these costs were recouped within four years in increases in rateable values (Pyne, 1973, p.24). Once grant availability was restricted, more studies tended to point at the 'economic and environmental benefit for local economies' generated by improvement (Jenkin, 1985, p.16). Shucksmith (1983) argued that home improvements result in greater rateable values and therefore extra revenue for local authorities (Shucksmith, 1983, p.180), a finding supported by the earlier work of Pyne. The SWEPC estimated that £325,000 (at 1972 prices) was being spent annually on the improvement of second homes in the south-west region (above improvement grant expenditure) (SWEPC, 1975, p.18). Home improvements often generate employment for local builders (Shucksmith, 1983, p.179), but on the down side, it might be argued that local builders are in fact being diverted from the task of building new housing for local people. However, this argument seems a little weak; it is unlikely that builders would be prevented from undertaking major construction contracts because they were being kept busy re-tiling roofs or laying patios. On the other hand, concern over the use of building resources on second homes could grow if the number of purpose-built dwellings increases substantially (Downing and Dower, 1973, p.28). There has been very little empirical work on the extent of second home improvements undertaken by owners; one notable exception however, is the work of Jacobs (1972). Jacobs' study of second homes in Denbighshire showed that owners go to a 'great deal of time and effort to improve their property' and only 13 per cent had carried out no repairs at all (Jacobs, 1972, p.21). The sums of money spent on repairs were considerable, with half of the owners studied by Jacobs having spent in excess of £200 (at 1972 prices). 11 per cent had spent more that £1,500 (Jacobs, 1972, p.22) and in aggregate, it was estimated that £314,096 had been spent on repairs across Denbighshire between 1965 and 1972, with many local builders benefiting from improvement work. The prior studies show that there is clear potential for the rural economy to benefit from improvement expenditure with the creation of employment in the construction industry. However, these studies are now somewhat dated and refer to a phase in second home growth when the supply of surplus rural dwellings (in need of substantial repair) in many areas was able to accommodate second home demand. This situation has now changed (see earlier discussion) and the types of properties used as second homes are now either unlikely to require substantial improvement work (because they are drawn from mainstream housing) or are new-built. Depending on the mainstream housing circumstances in the local area, new-built second homes may represent either an unwarranted drain on local construction resources or a welcome source of additional building contracts.

General Expenditure

General expenditure by second home users (on local goods and services) is a subject of much speculation and some empirical analysis. That owners bring additional revenue into the local economy during the tourist season (and at other specific times throughout the year) is a key argument in the case for the defence of second homes. Clearly, the growth in second home ownership provides an opportunity for keeping domestic holiday expenditure within this country, potentially assisting the balance of payments. At the local level, second homes provide a flow of money into the importing region, supporting the rural economy. Once an owner has acquired a second home, he/she (or friends and relatives) will visit the same area year after year without the need for any words of encouragement from a tourist office (Dower, 1977, p.157). Downing and Dower have argued that the annual influx of second home owners into importing regions represents a flow of permanence which is not characteristic of any other form of tourism (Downing and Dower, 1973, p.29). Opponents, however, contend that the general revenue contribution is slight. The most comprehensive study of this particular economic impact was undertaken by the South West Economic Planning Council in 1975. The general expenditure contribution in the importing region is a function of second home use (that is, the average annual length of time that a second home is occupied) and whether visits are of a long or short duration (that is, short being for a weekend, whilst longer visits might be in excess of a week).

During short visits, consumables might well be purchased in the exporting region; however, with longer stays in the second home, it will become increasingly likely that consumable goods will be purchased locally. Jacobs (1972) found that on shorter breaks in Denbighshire, there was some evidence of self-sufficiency, but local expenditure on consumables (e.g., milk, bread, groceries, meat and vegetables) increased on longer stays (Jacobs, 1972, p.26). The SWEPC study found that 32 per cent of owners only used their homes during the summer months (SWEPC, 1975, p.14) and that on average, 'the majority of second homes in the south-west are occupied for between a quarter and a third of the year'. The total amount of usage comprised visits by the owner and his immediate family, occasional loans to friends and relatives and in some areas, informal lettings on short tenancies to holiday makers during the height of the tourist season (SWEPC, 1975, p.16). In Denbighshire, Jacobs was able to demonstrate that the typical owner spends sixteen weekends a year is his second home and that many second homes were occupied during nearly every weekend during the summer months. In addition, 57 per cent of owners in Denbighshire lent their cottages to other people, giving an average of 17.6 weekends of use per annum; longer visits accounted for 7.5 weeks a year and therefore the use of second homes appeared to be intensive (Jacobs, 1972, p.19-20). Recurring general expenditure includes the cost of 'food' items, 'non-food' items (such as petrol or gas) and 'annual costs' (electricity and gas). The SWEPC found that £7 million was spent in the region annually on these recurring items (£3 million of which was spent in Cornwall alone). However, some owners already lived in the region whilst some consumables may have been purchased elsewhere. Allowing for a multiplier effect (using the 1.2 figure suggested by Lewes (1970)), the SWEPC estimated that second homes generated approximately £5.1 million of additional revenue for the south-west's economy each year (this figure included expenditure on improvements and initial purchase and represented less than 0.25 per cent of the region's income). The inevitable conclusion from this analysis was that the contribution of second homes to the *regional* economy was negligible. However, it was recognised that Cornwall took a disproportionately large slice of this additional income and that the *local* economic effect of second homes could be significant and should not, therefore, be discounted (SWEPC, 1975, p.21). In Caernarvonshire, Pyne showed that second home owners spent, on average, 83 days in the county each year and during this period, spent £1.2 million on food and non-food items (Pyne, 1973, p.25). The total expenditure in the county each year totalled some £2.8

43

million (including initial purchases and rates payments) and a study by the University College of North Wales in Bangor showed that this level of recurring expenditure was sustaining as many as 990 jobs and providing a reliable contribution to the local economy. In this particular analysis, Pyne argued that financial (and perhaps environmental) benefits from the second home phenomenon were visibly outweighing the costs although the economic up-side must be seen in the context of a significant socio-cultural down-side (see later discussion; Pyne, 1973, p.31). The prior studies tend to concede that the growth in second home ownership does have an economic benefit in importing regions and a significant element of this benefit relates to use and recurring general expenditure. However, it is also clear that the local impact is often far more important proportionally than the part owners play in the regional or national economy.

On balance however, even at the local level, the economic contribution of temporary residents must be set against the potential loss of contribution from full-time residents. If a dwelling changes use (from first to second home) and a local person is displaced, as might be the case in some areas where the demand for second homes is focused on mainstream housing, will the contribution of the second home owner be greater than the potential contribution of the displaced local? This point is raised by Shucksmith (1983) who argues that calculations of economic contribution 'fail to consider the alternative use of the housing and any associated expenditure thus foregone' (Shucksmith, 1983, p.180). He adds that in the Lake District, for instance, 'it may be that a second home owner has displaced a potential permanent resident who would surely have spent more than £480 (1972 prices) in the year' (Shucksmith, 1983, p.180). Again, the type of housing used as second homes in a specific area is a crucial issue; where surplus housing is brought into second home use, no displacement of the local population occurs and therefore, economic contribution need only be set against economic loss (of the type described by Shucksmith) where there is evidence of cross-competition for housing and subsequent displacement. At the regional level, different areas and communities will vary in their experience of the second home phenomenon in this respect as the balance of effective and non-effective housing stock is dependent on an array of factors including the past history of economic decline and dwelling abandonment, private and public sector new-build and public sector sales policies. In terms of economic impact, the potential adverse effect that second homes have on some rural communities mean that it is extremely difficult to incorporate second homes into classical models of domestic tourism expenditure. For example, the model by Archer (1973) (below) may be applied to second homes (as a type of tourist accommodation) but makes no allowance for the type of potential losses described above of the loss in income incurred by local hoteliers or guest house owners.

$$T = \sum_{i=1}^{N} \left(\frac{Bi \quad Li \quad Oi \quad Ei}{Si} \right) + Rc$$

Where 'T' is the regional tourism expenditure, 'i' is each type of tourist accommodation, 'B' is the number of bed-spaces (i.e., the number of visitors that can be accommodated each day), 'L' is the length of the season, 'O' is the average occupancy rate. 'E' is the amount spent by a tourist, 'S' is the average length of stay of a tourist and 'Rc' are caravan parking fees paid by outside owners of caravans (including static caravans) to site operators in the importing region. Numerous commentators have pointed out that recurrent expenditure by second home owners in the importing regions is extremely difficult to quantify; the effect of differing use patterns, self-sufficiency, potential loss of hotel revenue and the displacement of permanent residents (with the loss of their economic contribution) all vary between second home regions and between different local communities. However, analysis by the University College of North Wales in Bangor (Jacobs, 1972, p.50) appeared to offer strong evidence that localised economic impacts across rural Wales was generating £4.2 million annually for the Welsh economy as a whole and sustaining 1500 jobs. Similar analyses across Wales (see Pyne, 1973) led Jacobs to conclude that 'second home ownership brings large financial benefits to the rural areas' (Jacobs, 1972, p.50). However, the economic benefits in Denbighshire were largely welcomed because of an observed lack of direct competition for mainstream housing between local and newcomers. Jacobs noted that 'fact and local opinion combine to suggest that local families and second home owners do not compete in the same housing market. They seek different qualities from their housing to match their different purpose' (Jacobs, 1972, p.50). It is clear from the prior studies that second home growth brings economic gains in some areas which are normally characterised by the presence of second homes outside the effective housing stock. Measures to maximise the economic contribution from second homes must seek to divert attention from mainstream housing; a move which would bring positive economic and socio-cultural benefits. The separation of first and second home markets (if possible) would also allow for the accurate modelling and quantification of their economic impact.

Rates or Council Tax Contribution

In most of the prior second home studies, the economic contribution from rates (or council tax today) is often studied as a separate economic component and not aggregated with general recurrent expenditure. This is because rateable income is more directly linked to the sustainability of local services and the viability of rural communities. It was generally recognised that rates formed a significant part of the overall economic contribution of second home owners and two views on the importance of rates emerged. First, rateable income for the local authority helps sustain local services particularly as absent ratepayers (that is, the second home owners) place little strain on these services for much of the year. Second, the opposing view is that although these ratepayers contribute to the running costs of buses or post offices, their displacement of the permanent population results in only residual service demand for much of the year and a subsequent decrease in supply. If services are not used, then the local authority will not provide them because regular *use* rather than rates provides by far the greater part of service support income. The argument that the payment of rates (and increases in rateable values through the improvement of poor-quality dwellings) on the part of second home owners benefits the full-time local population has been used by Shucksmith (1983), Tuck (1973) and Bollom (1978). Shucksmith argues that because second home owners make little use of services compared to their rates contribution, they are in fact 'subsidising' local use of services such as schools (Shucksmith, 1983, p.180), but he concedes that the irony is that these same services may be withdrawn because of the same underuse which is in fact partially funding them. Other observers (Jenkin, 1985; Dart, 1977) also note that these rates must be used to cover the costs of new infrastructure (particularly where new-built second homes are concerned) and it may be some time before the local population experiences the benefits of increased local authority revenue. It is also clear that whilst the payment of rates (or council tax) may outweigh the *current* use of services, the second home owner:

> ... may, of course, make greater claims on those services if he later uses the property as a retirement home: some local authorities have expressed concern about the isolated nature of many second homes for this reason (Downing and Dower, 1973, p.31).

There is the potential in some areas for past rates contributions to seem meagre in the context of over-use of social services on the part of retirement migrants. However, it must be remembered that the second home phenomenon has not been exclusively responsible for the increasing numbers of elderly people living in the countryside. In fact, Davies and O'Farrell (1981) have pointed out that a greater appreciation of the difficulties associated with rural living (on the part of second home owners) may in fact discourage this group from retiring to the countryside. It could be argued that those people without this same experience and who are 'taken in' by the myths of the rural idyll as retirement approaches are more likely to settle in distant rural areas (perhaps in areas where they have spent one or two summer vacations in the past) and subsequently put pressure on already stretched rural services. Again, the economic impacts associated with rates or council tax will differ between local areas. Where second homes are predominantly new-built or conversions from surplus housing stock then there is a strong case for suggesting that second home owners contribute a 'hidden subsidy' to permanent residents through local authority rates or council tax (if the argument is up-dated). In Denbighshire for example, Jacobs was able to argue that 'the increased demand for public services is slight and the county and the district benefit substantially from the presence of the second home owners' (Jacobs, 1972, p.49). In contrast, where demand for second homes is accommodated in the mainstream housing stock, it can be argued that there is no revenue advantage; rates payments are simply replacing those that would have been paid by permanent residents and this displacement will mean that underused services are no longer offered. Similarly, these rates contributions will not cover the costs of council building programmes required to house those locals who find themselves excluded from the private property market.

The Overall Contribution

It is extremely difficult to provide an overall statement on the economic impact of rural second homes. The economic contribution of the rural tourist industry generally is difficult to assess given the lack of empirical evidence (Hoggart, et al, 1995, p.179). In the past, much of the work undertaken by local authorities has tended to conclude that second homes offer significant local economic benefits. One of the most positive contributions to this debate was offered by Jacobs (1972) who argued that 'all these [economic] benefits accrue without imposing any extra burden upon the local housing market' (Jacobs, 1972, p.50). However, this view is based on a number of highly subjective judgements regarding the economic costs and benefits of second homes which appear to defy rigid econometric analysis (Thompson, 1977, p.10). It appears that a key element in the local economic impact of second homes is the degree of separation between the first and second home property markets. Where second home demand is accommodated outside the effective local housing stock, the adverse impact of house price inflation will be largely avoided whilst money spent on general improvement may inject life into the local construction industry (the same may be true of purpose-built second homes). Likewise, the benefits of recurrent general expenditure or increases in council tax revenue will not be off-set by a reduction in demand, and the subsequent closure, of local services. In contrast, where there is no separation between the first and second home markets, competition between locals and outsiders for mainstream housing (the 'effective' stock) can generate inflationary pressure on property values. The subsequent displacement of former local residents will mean decreased demand for services despite the fact that local rateable income is sustained. On top of this, fewer of these properties will be in need of renovation work and therefore there will be little additional work for local builders. Clearly, the economic impact of second homes needs to be seen in the local context and is heavily dependant on the types of dwellings converted into second home use. The transitional phase in second home growth mentioned earlier (following the saturation of the non-effective housing stock and preceding policies to divert pressure from mainstream housing) is perhaps the most problematic in terms of economic and social impact. The identification of those areas at most risk requires careful analysis of local housing markets; without such analysis, it is impossible to draw any objective conclusions concerning the relative economic benefits of second homes. Past research in Ireland, however, has shown that given favourable conditions:

> ... tourists bring numbers, money and reassurance to those who live in isolated rural communities, with an enhancement to self-esteem perhaps having the most profound effect (Hoggart, *et al*, 1995, p.179; Messenger, 1969; Brody, 1973).

5 Environmental and Social Impacts

Compared to the other aspects of second home growth (economic and social), little work has been undertaken on the environmental and social impacts of second homes despite Coppock's assertion that interest in the growth phenomenon was at least partially driven by growing environmental and cultural awareness in the 1960s and 1970s. Part of the reason is that these concerns focus on new-build, the subject of more general development literature that deals with environmental issues under the auspices of sustainable development. Notwithstanding this, various references to environmental, social and cultural concerns are made in a number of the prior studies and it is these that are considered in this Chapter.

The Range of Environmental Impacts

The environmental impacts of second homes usually fit into one of three categories. The first of these is the impact of growth in the leisure industry generally, involving increasing pressure on environmentally sensitive areas (particularly National Parks and Areas of Outstanding Natural Beauty) and traffic congestion associated with an increasing number of people travelling into importing regions annually (Rogers, 1977, p.98). Downing and Dower (1973) have argued that two-home households use more building resources, more road space and are generally higher level consumers; this means that 'the prospect of widespread ownership of second homes does raise issues of resource consumption and environmental impact which are currently growing in the world's mind' (Downing and Dower, 1973, p.28). Secondly, there is the concern over increased development pressure in the countryside as the demand for new-build second homes rises and thirdly, there is the environmental and landscape (aesthetic) impact that restoration and conversion work has on the rural housing stock. In addition to these three main categories, there are also the environmental concerns that second home owners take with them into the importing region; these may have a positive impact, setting new environmental standards in the countryside or a negative impact, leading to protectionist attitudes at the expense of rural economic vitality. This latter aspect was examined in Belgium by Albarre (1977) who argued that whilst social conflicts might develop where second home owners adopt protectionist stances (he cites an example of owners objecting to pig farm developments in Wallonia because of the smell (Albarre, 1977, p.141)), a concern for rural conservation which recognises that the countryside is a 'working landscape' can help promote 'coexistence between various social groups' and halt the 'ineluctable breakdown in the rural environment' (Albarre, 1977, p.145). Psyhogios (1980) observed in his study of Nea Makri (Greece) that 'second home users, and mainly owners, once they have acquired their property they do care about the environment and sometimes take an active part in the preservation of the area' (Psyhogios, 1980, p.147). These types of issues are addressed in more general literature focusing on the sociology of community in which second home owners are viewed as just one group in a range of local and incomer consumption classes. The more general conflicts and social impacts surrounding second home ownership are examined later in this Chapter.

New Development and Conversion

In the prior studies of second homes, the main concerns have been the impact of new development and conversion. Conversion may offer positive environmental benefits as restoration and change of use can lead to the preservation of derelict properties on the verge of being demolished (Dower, 1977, p.156; Hoggart and Buller, 1995). Pyne (1973) argued that renovation contributes to the visual character of the surrounding area and noted that second home owners are often willing to accept advice on how renovations should be undertaken (Pyne, 1973, p.28). Downing and Dower (1973) have noted that preservation justifies renovation particularly where buildings have an historic value (Downing and Dower, 1973, p.32) and have argued that for this reason, public bodies might steer second home seekers to derelict or neglected buildings as 'such action could well have great value in relation to buildings of recognised historic and architectural interest both in town and countryside, and also to many humbler buildings of vernacular character' (Downing and Dower, 1973, p.37). Newly-built second homes are a different matter, having a parallel impact on the environment as other forms of new development. In this case, it is the responsibility of local planning departments to minimise environmental impacts. This applies equally to static caravans which require careful siting and stringent regulation, particularly where caravans are to be substituted with permanent chalet development. On the Lleyn peninsular in North Wales, however, Pyne (1973) argued that some new second home development was taking on a 'suburban' rather than a rural character and was having a detrimental effect on the rural landscape (Pyne, 1973, p.28). It appears from the prior studies that there is a clear divide between that second home development which has the potential to bring environmental benefits and that which may, if not carefully regulated, cause serious and probably irreversible damage. The short-cut conclusion is that conversion is potentially good and new-build is potentially bad. But these divisions are not universally applicable. In Scotland, for instance, the Dart study (1977) claimed that second home development could be either 'neutral or beneficial'; again the benefits accrue from the restoration of derelict properties, but the effect of new-build can invariably be neutral where good planning practices are applied; rich tree cover in Scotland meant that 'there was no need for development [of new built second homes] to be either obtrusive or alien to local character' (Dart, 1977, p.62).

Like the economic concerns, the environmental aspect of second home growth is closely linked to the types of dwellings used for this purpose. The trend in recent years has been away from derelict and empty surplus dwellings towards new-build and the use of mainstream housing stock. In the latter case, the environmental impact is likely to be neutral. In the case of new-build, the impact will range between being neutral and being negative depending on the effectiveness of local planning control and the application of design and siting standards (Downing and Dower argue that new-build developments should be directed towards disused or reclaimed land as development of this nature would have a lesser environmental impact and could potentially attract European Community grants; Downing and Dower, 1973, p.37). In France, Hoggart and Buller (1995) have argued convincingly that second home acquisitions are having a positive environmental benefit with British buyers 'making significant additions to the rural housing stock of France through the acquisition and renovation of buildings that are unsuitable for human habitation' (Hoggart and Buller, 1995, p.188). However, the rural housing market in France can still accommodate this type of demand in surplus stock; the same is not true in Britain where the outside demand for *effective* housing in some importing areas has generated economic, environmental and social tension.

Social Impacts

Perhaps the most controversial issue surrounding the expansion in second home ownership in recent years has been the social effect that this growth has had on importing regions and on particular rural communities. Supposed social impact is closely related to the negative economic effects that second homes bring in some circumstances, notably, inflationary pressure on house prices, the displacement of permanent residents and the cessation of community services. Where these processes are rightly, or wrongly, attributed to a growing number of second homes, opposition is often mobilised against 'middle-class incomers with agnostic views and little appreciation of rural life'. When the divisions are drawn, it appears that a complex housing market has been reduced to just two competing groups; working-class, low income locals and middle-class, high salaried incomers. In this context, it is easy to attribute blame for various social tensions. In Wales, this dichotomy is supported by cleavages running along cultural lines. Emmett (1964), for example has argued that 'in the presence of the enemy [the English], Welshness is the primary value; deacon and drunkard are friends, old schisms become unimportant' (Emmett, 1964, p.13). Emmett adds that in Wales, the English take the place of the upper classes and 'nationalism is the dress in which class antagonisms are expressed' (Emmett, 1964, p.23). However, this is not the way in which rural *housing markets* are constructed and before considering the social impact of second homes, it is necessary to provide a more realistic framework in which a range of different groups compete for housing.

Competing Housing Groups

Shucksmith (1990b) argues that housing class, in the Weberian tradition, can be seen as a function of market power and therefore the rural property market has clear implications for the class structure of rural societies (Shucksmith, 1990b, p.210). However, the nature of the market and particular tenure divisions mean that there are no simple dichotomies between rich and poor or locals and newcomers. Saunders (1984), for example, argues that owner occupiers form a middle 'domestic property class' who are advantaged by the cumulative potential of their tenure whilst those in rented accommodation have no similar advantage. Clearly then, the effect that outside competition has on a rural market will depend on the level of indigenous owner-occupation and potential cumulative wealth (determined by local property values). Similar classifications have been developed by Ambrose (1974), Pahl (1966) and Dunn, Rawson and Rogers (1981). Domestic property classes are characterised by 'consumption cleavages' which define access chances and the ability of different households to compete in the housing market. Shucksmith (1990b) contends that access chances broadly differ between households with 'low income and low wealth' and 'more prosperous groups' (Shucksmith, 1990b, p.225). However, there are numerous sub-divisions within theses broad categories. The first group will include young couples and single persons barred from entry to waiting lists (this group is often perceived as experiencing the greatest difficulties in gaining entry to the housing market; Cloke, *et al*, 1994, p.4), other tenants of rented or tied accommodation, pensioners retired from local employment or former in-migrants now facing financial difficulties and local authority tenants who are more fortunate and who may be eligible to buy their home with a discount. The latter, more prosperous group, might include indigenous owner occupiers, retirement migrants with capital available from a previous home, holiday home buyers and commuters. Clearly, this typology demonstrates that certain groups will face disadvantage in the rural housing market, not only in competition with newcomers but also with indigenous owner-occupiers who have the advantage of an established equity base. Shucksmith concedes that the above framework is empirically derived and will not have universal application; he notes, however, that all examinations of competing housing groups must recognise the importance of income and wealth, present tenure, life-cycle status and the motives for residing in the rural area in determining housing outcomes. Rural housing markets are more than just local interest versus second home demands.

Social Impacts, Social Groups and Attitudinal Research

The growth in second home ownership in rural England and Wales has generated both a realised social impact and an emotional reaction. The realised element has already been touched upon and includes the displacement (and subsequent replacement) of some rural residents (i.e., those low income, low wealth groups unable to compete in the property market) and the disruption of rural services as regular demand declines. This realised element might include the erosion of the socio-cultural fabric of receiving areas as rural values (and language in Wales) are displaced by

53

the urban values of incomers. These elements may be followed by emotional reactions which are far more difficult to gauge, but which are usually expressed in terms of 'attitudes' towards second homes and their owners. According to Thompson (1977) the social impact that second homes have on rural communities (including community decline) may be judged directly against changing local attitudes (that is, attitudes are a function of social change and impact). The importance of attitudes and attitudinal research was certainly accepted by Bollom (1978), but it has long been acknowledged that judging attitudes and quantifying 'social intangibles' may prove extremely problematic (Edwards, 1957; Osgood, Tannenbaum and Suci, 1958; Triandis, 1970; Summers, 1970; Oppenheim, 1966). Edwards (1957) argues that feelings on subjects, particularly controversial ones, may not be easily articulated and:

> ... we may, for example, have both positive and negative affect associated with the same psychological object. How then are we to weigh and evaluate the strength or intensity of the two of the opposed effects and decide whether we like or dislike the object? (Edwards, 1957).

Commonly, it is supposed that attitudes are constructed of three core components. The first is the 'cognitive' component which relates to the experiential background behind attitude formulation and may, for example, involve direct experience of second homes, either as a local vendor accumulating profit from a sale or as a displaced local. The second is the 'effective' component which refers to the emotions which shape ideas (in Wales, an emotional attachment to language may determine reactions to those elements of social change which affect the survival or use of the language). The final component is 'behavioural' and determines a pre-disposition to action, which may produce hostility towards second homes expressed in the form of protest. Oppenheim (1966) explains that attitudes are closely related to behaviour, although this relationship may not be constant over time. Thompson (1977) however, argues that the growth in second home ownership and the realised social impacts are generating a sustained set of attitudes and social responses (Thompson, 1977, p.18). The pressing issue, therefore is how these attitudes, which are indicative of social change, should be measured. The simplest approach is to derive linear scales relating to attitude statements (concerning particular concepts) and questioning respondents directly (i.e., on the lines of 'do you think second homes are: good, acceptable, not acceptable or bad?'). A less direct approach used by Thompson (1977) was to employ 'semantic differentials' (Osgood, et al, 1958) which measure reactions to stimulus concepts on three scales and attempt to conceive the process of judgement, how meaning can differ between individuals and the semantic space around a concept (the degree to which individuals apply different meanings to an object). Using the semantic differential method, Thompson was able to derive 'D' scores relating to attitudes towards different phenomena (e.g., second homes, tourism, the Welsh language etc.). When he compared the 'attitudes' of second home owners with those of locals, he found that whilst locals reacted more favourably to concepts such as 'local people', 'the chapel', 'Welshness' and the 'Welsh language', second home owners tended to gravitate towards concepts of 'Englishness', 'Second homes' (and their owners) and 'tourism' (Thompson, 1977, p.35). Although attitudes may be indicative of social impact (of a particular process or component of change), it is difficult to define the nature of any relationship. On the one hand, it might be hypothesised that increasing second home density generates greater impact (in terms of the displacement of population and the disruption of services) and is therefore matched by increasing polarisation of local people's attitudes. On the other hand, the decline of community may be matched by the erosion of social norms and perhaps a mellowing of attitudes over time; in this context, the acceptance of second homes may be seen as a later feature of community decline (Thompson, 1977, p.39).

One of the most important prior studies on the social impact of second home growth (defined in terms of changing attitudes) was conducted in North Wales by Chris Bollom in 1978. In five case study areas (Penmachno, Cwm Penmachno, Rhiw, Croesor and Llansannon), Bollom examined the *density* of second homes and the effect that higher density has on community life and the attitudes of local residents. Hostility and opposition to second home growth is often a function of local leadership, the vitality of local village life and the influence of traditional institutions which may act to politicise opposition. It follows that where opposition is vociferous, second homes have not, as yet, had a great negative social impact and these areas may be characterised by either low densities of second home ownership or patterns of ownership (concentrated in surplus housing or in new built developments) which are not causing serious social disruption. Elsewhere however, high second home densities or the accommodating of demand in mainstream housing may lead to the displacement of local residents and the submergence of traditional cultural institutions (such as chapels in Wales). Paradoxically, in those areas experiencing the greatest social impact, submergence of institutions (which often form opposition structures) may act to alleviate the tension between groups and these communities are eventually characterised by accommodation and acceptance (Bollom, 1978, p.vii).

One way of gauging social impact therefore is to draw an inverse relationship with opposition; vocal opposition is the hall-mark of the healthy rural community whilst grudging acceptance characterises those communities where social institutions have declined in the face of population displacement. Bollom argues that this relationship occurs because in those areas little affected by second homes, people more readily identify with traditional life-styles and react against what they see as the potential threat posed by second home growth. For this reason, attitudes may become less favourable than those typical of people in high density areas where the prevailing attitude amongst the remaining elderly population (those perhaps least able and most entitled to object) is one of accommodation (Bollom, 1978, p.113). The most potent hostility therefore does not always materialise in those areas experiencing greatest social impact; rather, hostility and opposition occurs where local voluntary organisations give shape to the suspicious attitudes of local residents. Bollom was able to use this interpretation to shed light on that element of hostility in the five case study areas that could not be explained in terms of realised social impact.

Clearly, in terms of realised social impact, the density of second home growth (and the types of dwellings accommodating demand) is still important as population must be displaced before opposition structures faulter and eventually decline. It follows that the social impact of second homes will differ between settlements of different sizes. Larger absolute populations are arguably at less risk from the encroachment of second home owners and, for this reason, growth of second home ownership in small communities, lacking opposition structures, is often seen as the greatest problem. Although Bollom does not address the specific issue of different second home *types* having different impacts (social, environmental and economic), he notes that a lack of 'locational separation' between original and new residents is a key factor in determining patterns of opposition. He argues that where separation does occur (such as in commuter villages in the English south-east), people become more attuned to notions of national class divisions because newcomers are more easily identified and grouped together (see Pahl, 1970). In this context, second home estates would not be viewed as monuments to economic decline. In their existing form (occupying mainstream housing), however, they are viewed as just that and opposition is readily mobilised against them (Bollom, 1977, p.119). Evidently, a feature common to both the realised and emotional social impact of second homes is the fact that it is only second homes accommodated in the effective housing stock which have the potential to amplify community decline (which was initially tied to the underlying patterns of economic decline in the importing region) and generate lasting social tensions and conflicts.

In the majority of the more general second home studies, reference to social impact is implicitly related to competition for mainstream housing and the replacement of permanent residents. Jenkin (1985), for example, argues that second homes have escalated the problem of community stagnation and is critical of some studies (e.g., Bielckus, *et al*, 1972) which have suggested that second homes have not adversely affected host communities. Clearly, it would be dangerous to field this argument in culturally sensitive areas if the studies do not distinguish second home *types* and their contrasting socio-economic impacts. Jacobs (1972) has argued that the second home issue is perhaps of greater social importance in Wales than in England where, at least until recently, 'regional cultures have practically ceased to exist' (Jacobs, 1972, p.33). Although it is possible that conflicts might be more speedily resolved in parts of Britain which escape the problems of language, it is clear that Jacobs' 'lack of regional culture' argument for England is without foundation. Whilst the displacement of permanent residents might represent a principal element in the Anglicisation of some Welsh communities, the acquisition and renovation of isolated dwellings outside the effective housing stock will not have a similar effect. However, opposition will focus on such dwellings if their impact is confused with that of mainstream housing purchased by second home seekers.

Community Unease and Socio-Linguistic Concerns

Between 1979 and 1990, the extremist group, '*Meibion Glyndôr*' carried out 171 arson attacks on such properties in North Wales (Pilkington, 1990, p.18), returning them to the derelict state that many were in prior to renovation by second home owners. It has been claimed that the threat of being 'burned out' resulted in an 'English retreat' from North Wales; across the county of Gwynedd, the total number of second homes fell by 2,500 to 8,000 by 1994 (*The Observer*, 17 April 1994), although it is difficult to isolate the effect of the arson canpaign from other factors such as the collapse in the property market in the early 1990s or the re-focusing of some demand on properties abroad. These attacks were motivated by fundamental misconceptions, the ease with which such isolated dwellings could be targeted, and a desire to focus publicity on the entire second home issue. In Wales, second homes (*per se*) have come to symbolise a threat to Wales' distinctive culture and language making it particularly difficult to disentangle the various strands of the second home debate (that is, the underlying causes of social decline and the differing social and economic impacts of different second home types). Whilst it is clear that in some circumstances, 'the strength of the chapel and the vitality of the language may be threatened' (Shucksmith, 1983, p.181) directly by second homes, Newby (1980a) has argued that 'loss of community' is the result of more general socio-economic change and cannot be attributed to newcomers (Newby, 1980a, pp.164-178). On the whole, second homes are tangible and convenient scapegoats whilst the less tangible underlying causes of 'community decline' is the collapse of the economic base and the changing nature of family life and other social networks. Arguably, rural communities, by becoming less 'traditional', are becoming less recognisable and a fear of change is generating a hostility towards those features of the rural landscape which have come to symbolise that change. However, these more theoretical and holistic perspectives do not disguise the fact that an influx of 'wealthy town-dwellers with cultivated accents and liberal or agnostic views' is often accompanied by a degree of cultural shock (Downing and Dower, 1973, p.30) in the summer months, followed by a loss of social vitality in the winter and the creation of 'ghost' communities. This shock may be emphasised by the 'physical configuration of housing, which often segregates the local working classes in social housing estates, while incomers colonise the old core of the village' (Hoggart, *et al*, 1995, p.216).

The concentration of second homes in Wales is central to the perception of a general "problem". Local reactions to newly built *holiday home* developments are well documented – but although they may represent a blot on the landscape, their social impact is less obvious. Holiday villages remain "ghost" communities for much of the year, occupied only during vacation periods. On the other hand, second homes *within* village cores "rob" locals of the dwellings which should "rightfully" have been their's: it is external sources of core housing demand which is perceived to represent the greater threat. Philip O'Connor's seminal work on second homes in 1962 highlighted local despondency over the plight of the language in Croesor in the light of new English monoglot arrivals; the erosion of Welsh usage was seen to herald a loss of community (O'Connor, 1962). In the 1970s and into the 1980s, a proportional decline in the number of Welsh language speakers in the traditional language "strongholds" was attributed to counter-urbanisation alongside the broader impacts of tourism (Bowen and Carter, 1975), retirement related in-migration and the growth in second home ownership (Aitchison and Carter, 1985; 1986). However, explanations of language change and stability have, to date, only defined a loose association between declining Welsh usage and second homes. This association stems from the fact that incomers from a dominant culture are apparently helping shift the linguistic balance in the traditional strongholds of the Welsh language (James and Williams, 1997) – and contributing to an eroded sense of community.

However, as previously stated, we should avoid confusing a "loss" of community with the broader processes of social change. An eroded sense of community may have as much to do with changing local values as any external influences. Similarly, the encroachment of newcomer groups may only appear significant where there is a net outflow of younger households (moving to seek better employment or educational opportunities) and this outflow, may in itself, create the market openings which are exploited by second home buyers. Indeed, Champion and Townsend (1990) have argued that 'second home use may yield net benefits to the local economy where they occupy premises that would otherwise be vacant, due to out-migration etc.' (Champion and Townsend, 1990). They do not, however, state the nature of these benefits.

Economic common sense suggests that there is a potential for second home purchases to cause an inflation of house prices that may then affect the housing opportunities facing some local households. This "common sense" perspective has shaped perceptions of a second home "problem". However, given that there is a real shortfall in housing supply in the countryside, and that this may undermine the stability of local communities, there is a pressing need to look for ways of enhancing the housing opportunities of local buyers. Where necessary, this may involve discouraging those conflicts of interest that may be derived from particular types of second home purchases.

Conclusions

Bielckus, et al argued almost 30 years ago that more research was required on the changes second home owners bring to rural communities from a sociological viewpoint (Bielckus, et al, 1972, p.145). Prior research has tended to concentrate on the realised social impacts, usually expressed in terms of population displacement and the disruption of services (including traditional institutions such as chapels in Wales) alongside attitudinal responses which are seen as being in some way indicative of deeper-rooted social change. Bollom has shown that community opposition to second homes is not a reflection of community decline but rather the strength of social structures and the vitality of village life. If this is the case and second homes can have a detrimental social impact, then alarm bells should only be sounded when opposition to this component of social change ceases. Perhaps the most important point to take forward in this discussion is the fact that the social, economic and environmental impacts of different types of second home development differ markedly. Whilst the renovation of surplus stock may act as a positive economic and environmental force in the countryside, the *change of use* of mainstream housing stock may be seen as the natural successor to economic stagnation as a force contributing to the social decline of rural communities (see Figure 1 overleaf; the 'tendency in some regions is for second home and new in-migrant populations to construct new properties in their chosen destinations' (Hoggart, *et al*, 1995, p.181; Weatherly, 1982; Bontron, 1989, pp.88-92).

Figure 1: Comparison of demand types and associated costs and benefits

Surplus dwellings	Mainstream Stock	Purpose Built
	United Kingdom	
	Scandinavia	
	France	
	Spain	
	Greece	
Benefits		
May enhance the quality of the landscape and bring improvements to the rural housing stock. There will also be work for local contractors, an economic input (through new tax revenue, local expenditure) and few associated social problems	Displacement of local economic input is likely to outweigh the new input from second home owners	Purpose built second homes are subject to normal planning controls and when grouped together and properly sited, may form part of a wider economic strategy, generating new income and employment
Costs		
There will be few associated costs providing these dwellings are surplus to local need	New competitors in the housing market mean rising house prices, the possible exclusion of local buyers, a reduction in rented accommodation, a cessation of local services, and social tension	May bring environmental costs if poorly-planned (particularly if developments are too large) and may conflict with other recreation needs (ie in Scandinavia)
Policy Responses		
Encourage by pro-actively seeking buyers or reinstating grant rights	Change of use planning controls or property market licensing	Encourage but regulate through normal planning procedures

PART II

POLICY AND PRACTICAL RESPONSES IN EUROPE AND THE UNITED KINGDOM

6   European Perspectives

Introduction

In the first Chapter it was noted that second home growth in Europe was already a 'mature' social phenomenon before the 1970s. This fact led many British observers to draw comparisons and consider what 'lessons might be learnt' from experiences abroad. However, these domestic studies tended to recognise that:

> Although reference to foreign experience may provide guidelines, differences in social, cultural, economic and political life-styles are likely to invalidate any clear comparisons or conclusions (Williams, 1974, p.31).

Kemeny (1995) has argued that after the Second World War, 'Western Europe came under the cultural influence of the USA and its English-speaking allies' (Kemeny, 1995, p.3) leading to an 'unconscious assimilation of political and ideological perspectives'. In contemporary research, this assimilation may be manifest as a fundamental misunderstanding of the social and political processes at work in neighbouring European countries. From an Anglo-centric perspective, it is all too easy to commit what Kemeny terms a 'Romeo error' (referring to the mistaken diagnosis of death in the Shakespearean tragedy), applying the same interpretations used in Britain to processes observed abroad (Allen, *et al.*, 1999). In this study, no attempt is made to draw direct comparisons with the British situation; rather, the purpose is to consider the range of European second home experiences outside the UK, to highlight the observed differences and also to examine how countries have responded to *their own* problems. The UK situation is considered in detail in Chapters Seven and Eight.

Facts Through Figures

The problems of collating data concerning the number of second homes in the UK pale in comparison to the difficulties of constructing a Europe-wide picture. The European Union has long been interested in formulating new methods to collect data in the 'field of tourism' in order to combat this problem (Official Journal, c236, 1995, p.20). In the early 1970s, this lack of data was particularly acute.

Table 1 The ownership of second homes across Europe (1970-1988)

Country	% of households with a second home		
	1970[1]	1980[2]	1988[3]
Sweden	22	nd	nd
Norway	17	nd	nd
Spain	17	nd	8
France	16	12	9
Portugal	10	nd	4
Denmark	10	10	12
Austria	8	nd	nd
Switzerland	8	nd	nd
Belgium	7	4	5
Finland	7	nd	nd
Luxembourg	6	nd	nd
Italy	5	4	6
West Germany[4]	3	nd	nd
Netherlands	3	2	1
United Kingdom	3	nd	3
Ireland	2	2	2
Greece	nd	nd	12

Notes: (1) *Source: Survey of Europe Today* (Reader's Digest, 1970); (2/3) Source: European Union (1996) *Social Portrait of Europe* (Luxembourg, Office for Official Publications of the European Communities), p.172; (4) The EU had no retrospective data for unified Germany in 1980 and 1988; (nd) denotes 'no data available'.

However, in 1970, a survey of 20,000 households across Europe considered the issue of second home ownership and the results were published in the *Reader's Digest* 'Survey of Europe Today'. It was acknowledged at the time that the survey potentially overestimated the proportion of households with second homes in certain countries (Downing and Dower, 1973, p.20). At the same time, it was not specifically concerned with EU member-states. Notwithstanding this, the tabulation below compares that data to more recent data collated by the EU for the old twelve member states (these twelve have since been joined by Sweden, Austria and Finland).

Table 1 offers a fragmented picture of second home ownership across Europe. The apparent decline in ownership in some states (notably France and Spain) is likely to be due to overestimation of the phenomenon in 1970 and then underestimation in 1980 and 1988. The lack of data in Germany is largely a result of the administrative changes associated with re-unification. Similarly, the lack of official EU data for Sweden, Austria and Finland reflects their recent membership of the Union. What is clear, however, is that the UK (along with Ireland and the Netherlands) has maintained its marginal position in terms of second home ownership. However, the UK has the second highest birth rate (next to Ireland) and household formation rate in Europe (Eurostat, 1993, p.5) meaning that if the figures are to be believed, the rate of second home growth has kept pace with general household formation over the last twenty or so years. The 1988 figure precedes the height of the British housing boom in the late 1980s and therefore it is likely that it fails to reflect the increasing number of acquisitions from the summer of 1988 through to 1990.

These figures bear testimony to the unsatisfactory and inaccurate way in which some data are collected across the European member states. However, the main purpose of this chapter is not to paint a comprehensive statistical portrait, but rather to consider changing patterns of ownership (even if underlying processes prove elusive) in individual countries and ways in which responses may have been formulated. The list of countries examined below is not exhaustive; the descriptions draw heavily on the prior domestic studies already outlined alongside some academic papers which have been published more recently. Invariably, they draw on the English-language literature.

Belgium

Throughout the 1960s, Belgium experienced a substantial increase in the number of second homes. By the mid-1970s, there were estimated to be 150,000 rural cottages and in excess of 200,000 chalets built without planning permission. Particular concentrations were developing in the French-speaking area south of the Sambre (Albarre, 1977, p.139). Belgian analyses of second home ownership tend to include urban apartments used by businessmen (that is, *"pieds à terre"* (Clout, 1973, p.750)) and the overall quality of data varies between different Belgian regions. Albarre (1977) claimed that a particularly acute problem in Belgium was the deteriorating relationship between second home owners and local residents, especially farmers (Albarre, 1977, p.140). Brier (1970) highlighted some increases in house prices in certain second home areas which may have been encouraging the process of rural depopulation. Increases in communal expenditure (on roads, water and electricity supply) were also attributed to the growth in the number of second homes (Brier, 1970). Of particular concern, in terms of social conflict, was evidence that middle-class incomers were often objecting to certain farming practices (either because of the noise or the smell). In Wallonia, owners had a particular dislike for pig farm developments which corrupted their idealised countryside (Albarre, 1977, p.141).

In many areas, there was evidence of a growing social rift between the incomers and the host population. However, Albarre demonstrated that this rift was not universal and could be bridged if *common concerns* could be identified and subsequently used to bring together the new and the old population. The commune of Sivry on the French border covered an area of 2,300 hectares and had a resident population of 1,411 at the end of 1971. The undulating landscape was punctuated with small farms divided by distinctive hedgerows; approximately one in eight dwellings was used as a second home, a symptom of the rural depopulation observed in the area since the 1950s and the beginning of the steady decline in the agricultural economy. Throughout the 1960s, a number of urban households purchased vacant properties in Sivry, attracted by the area's rustic character. When the second home owners arrived, they took over not only the buildings but also the hedgerows and recognised the importance these had in maintaining the overall character of the landscape. Schemes for hedgerow conservation (including a competition for the best-kept hedgerow) brought locals and incomers together and promoted an awareness of the interest that incomers had in the rural environment amongst locals and particularly farmers.

Albarre claimed that through this process, it was possible to remove some of the obstacles to 'coexistence between various social groups' (Albarre, 1977, p.145). In effect, the incomers came to an area of depopulation (and subsequent neglect) and made an important and significant contribution. A communication channel was established between town and country (built on understanding) which demonstrated how people *can* live together. Albarre notes that:

> Geographers, aware of the increasingly complex interpretation between the urban and rural environments, have encouraged this encounter between farmers and second homers in order to help communication between them, for without this, there would surely be an ineluctable breakdown in the rural environment (Albarre, 1977, p.145).

Patterns of socio-economic change had produced a new social configuration in the Belgian countryside and the divisions between social groups were clearly visible. However, Albarre showed that it is possible to identify common ground and use this as an arena for overcoming social tension by fostering greater understanding. The problems posed by social change are often balanced by the new opportunities offered; it may be the task of both locals and incomers to decide which way the scales should be tipped.

Denmark

In 1970, there were estimated to be between 140,000 and 145,000 second homes in Denmark, mainly located in coastal areas (a pattern which contrasted markedly with that observed in Sweden and Norway where the majority of second home developments had occurred in the forested and mountainous areas). The parcelling of land for new developments since the 1960s had generated particular concern as public access to beaches was being limited. At this time, half of Denmark's second homes were located in Zealand, with demand coming predominantly from Copenhagen. The development of second homes in Denmark has created a number of planning problems in recent years. Sporadic development in the 1960s was acknowledged as having a detrimental environmental impact; in response, it was recognised that 'recreational' home developments (in groups of roughly 200 dwellings) could limit the impact on the countryside particularly if they were viewed as representing a settlement in their own right and all normal planning guidelines were adhered to. However, some of these 'settlements' had, in the past, been built too close to urban centres and over time, the growth of cities had engulfed these second homes causing a change of use and the creation of poor quality housing ghettos (Bielckus, 1977, p.37). In order to combat this problem, regional plans were used to specify minimum distances away from urban centres for the siting of second homes; similarly, planning guidance was strengthened in order to ensure that no second homes were without adequate water supply or sewage disposal facilities. The measures adopted by the Danish government in the 1960s and 1970s did not seek to limit second home development, rather they sought to ensure that the qualities which attract second homes to an area are not lost through poor planning. At this time, opposition to second home development was in its infancy and the arguments of the anti-development lobby were given little credence.

More recent Danish legislation has focused on the over-riding importance of securing a first home and maintaining the socio-economic viability of host communities. The Danish Government has legislated on both the use of residences and the acquisition of property. First, planning legislation distinguishes between permanent and 'secondary' accommodation. If a new dwelling was granted planning permission on the grounds that it was to be used as a 'permanent' residence, or if a property has been used as a permanent residence over the past five years, then permission must be obtained from the local commune authority before any *change of use* may be implemented. This rule applies for a range of 'use' changes and has been formulated as a response to a growing housing shortage in some parts of Denmark. Danish local authorities have a social responsibility for households seeking accommodation and the control they exert over change of use assists them in this task. The second legislative change relates to the acquisition of property on the part of persons not domiciled in Denmark. Essentially, non-Danish nationals need to apply for a permit from the Minister of Justice in order to acquire real estate. On Denmark's accession to the European Union following the Treaty of Rome, the law was amended to allow citizens of other European member states to acquire real estate with the intention of working and residing in Denmark. This right, however, was not extended to the acquisition of second homes, a point emphasised in the original treaty:

> Property in Denmark - Notwithstanding the provisions of this Treaty, Denmark may maintain the existing legislation on the acquisition of second homes (Treaty of Rome; Protocol 2).

In effect, foreigners must still apply for a permit if they wish to purchase a second home. Social concern has always been central to the second home issue in Denmark. In the 1960s, controls on second home development aimed to ensure and maintain general access to the countryside; the availability of recreational opportunities (to everyone and not just a privileged few) was seen as an important social objective. More recently, attention has focused on the availability of housing for permanent residents with both planning and housing law being used to ensure that housing shortages are not accentuated by second home demand. The Danish strategy in relation to second homes and local interests has been to accommodate outside demand, but only after wider social objectives (leisure and housing) have been achieved.

Finland

Williams (1974) has noted that in Finland there has been marked local interest in the second home phenomenon, particularly in the Åland Islands where the Finnish language (widely spoken throughout Finland) was becoming virtually unknown because of the presence of second home owners (Williams, 1974, p.36). Åland province has been an area of political controversy for many years and particular border disputes between Finland and Sweden centred on the province in the early part of this century. In the 1920s, the League of Nations assigned power of administration over the territory to Finland with certain conditions of decentralised control. For this reason, Åland has developed its own brand of provincial legislation which places great emphasis on the rights of 'natives' over outsiders (encapsulated in 'hembygdsrätt'). Åland's law of real estate acquisition, for example, regulates property rights in the province and aims to ensure that all land stays in the hands of people born in Åland. Those not fortunate to have 'hembygdsrätt' (local property rights) cannot acquire property without a permit and these are only granted to persons connected to Åland by family ties or employment needs. Even where permits are allocated, the property which can be acquired is restricted in type and size.

Clearly, these controls in Åland will restrict all types of property acquisition (including second homes). Similar measures have been implemented in other parts of the country including Northern Savo and are a reflection of the way in which provincial devolution has been administered (a British parallel might be the relationship that the mainland

71

shares with the Channel Islands). Apart from the provincial real estate legislation, there have also been national measures to control the growth of second homes. First, persons not domiciled in Finland for a continuous period of five years require a permit to acquire a holiday residence (whilst no such 'licensing' scheme extends to the purchase of first homes; a move which may have delayed Finland's recent accession to the EU). This particular housing market restriction will remain in force until a scheduled review in 1999 and is intended to curb competition between local people and second home seekers, reducing the potential for house price inflation. Second, legislation has also been implemented which controls the number of farms taken out of farming use. If a person wishes to acquire a farm on which to pursue agricultural activities, he is free to do so. If, on the other hand, the farm is to change use (e.g., the farming function is to cease or land is to be sold and the farm residence is to become a second home) then a permit will be required.

The provincial system of Government in Finland has allowed some areas to take quite drastic steps (or at least steps which appear drastic from a British perspective) in order to control the housing and property markets and prioritise local interests. More general controls have also been implemented which seek to ensure open access to first homes and protect the agricultural base where it remains viable. Clearly, however, the most important point here is that the way in which the property market is regulated is a function of peripheral culture and the power wielded by peripheral Government. Where power is devolved, the local or regional perspective is likely (in some instances) to take precedence over national concerns. Where this is not the case, as has been the case in England, Scotland and Wales until very recently) the national perspective prevails and the integrity of national markets is seen to be of greater importance than local interests.

France

Two phases of academic interest in the French second home phenomenon can be identified. The first is associated with the work of Clout in the late 1960s and 1970s and the domestic housing market. The second is more recent, focusing on Hoggart and Buller's concern with British property buyers in rural France during and after the British property boom of the late 1980s.

Clout (1977) notes that the acquisition of second homes in France has a long and varied history and one which follows a pronounced pattern of social democratisation. In pre-Revolutionary France, *châteaux* and country parks owned by the nobility were widespread; these were replaced in the nineteenth century by less grandiose weekend houses owned by rich provincials. Modern patterns of ownership took root in the inter-war and post-war periods and the new 'second homes' comprised country houses used for short periods, holiday homes or furnished flats used for tourist purposes (Clout, 1977, p.47). The French population census (which has collated data regarding second homes since 1962) along with 'Cadastral registers' and other taxation documents today provide the key sources of information regarding second homes across the Channel (Bonneau, 1973, pp.307-320). By 1970, there were 1.5 million second homes across France, although patterns of growth since 1945 have been complicated by the return to first home use of second homes in some northern and southern parts of the country. After the war, some vacation residences in Brittany and Normandy were returned to first home use in order to overcome a shortage of housing created by the devastation of the Allied invasion. In 1962, many southern second homes met the same fate as *pieds noirs* were repatriated from Algeria. Between 1963 and 1966, 25,800 second homes became primary residences, either because of the process of repatriation in the South or because they became overwhelmed by the rapid expansion of many large urban centres (Palatin, 1969, p.747-57).

Despite these losses, the dominant trend was still one of growth; the construction of new second homes more than doubled between 1963 and 1966 from 10,000 per annum to more than 24,000 (Clout, 1977, p.50). Two thirds of these purpose built homes were located on the coast with the remainder split equally between rural and mountain areas. Psyhogios (1980) has noted that in France:

> ... regional planning corporations have advocated the designation of areas for second home development on the basis that essential utilities could be provided far more cheaply than in dispersed locations and would create far less conflict with farming, forestry and countryside preservation interests (Psyhogios, 1980, p.63).

Hall (1973) has traced the way in which second home developments were 'steered' in the Languedoc-Rousillon area with the creation of new coastal resorts comprising various service provisions and new-built second homes (Hall, 1973, p.173-75). By 1967, 2.8 million households made use of 1.2 million second homes with 'ownership as opposed to use, [still being] most widespread among affluent strata of French society' (Clout, 1977, p.51). Many owners had relatively high annual incomes and tended to be engaged in professional occupations although in the late 1960s, Clout argued that this pattern was slowly changing (Clout, 1969). Many of the users were the children of older owners who were normally town dwellers (just 2.5 per cent of French second home owners at this time lived in rural cantons). One of the most important points to be raised in Clout's research at this time was the role of *inheritance* in the second home market. In France, it was commonplace for a city-dweller to inherit a rural home from grandparents (Clout, 1977, p.58) and although the national picture was far from clear, anecdotal evidence served to underline Clout's point. In Hérault, 45 per cent of owners had inherited their second home whilst the same was true of 25 per cent of owners in the Paris basin.

By the 1960s, the growth in the number of second homes across France had become the subject of some concern and considerable social comment. Barbier (1968) expounded the economic and social benefits that second homes could bring to importing regions including the local profits derived from sales, the generation of new employment for local craftsmen and the opportunity for rural people to widen their social contact. On top of this, second homes brought the opportunity to anchor dwindling rural populations through an injection of new money. As if by magic, second home owners are drawn to declining regions without the need for any strategic guidance. Cribier (1966), on the other hand, argued that the proliferation of second homes and the subsequent rises in house prices was, at best, producing unbalanced communities and, at worst, accelerating the process of depopulation (Cribier, 1966, pp.97-101). These same arguments became central to the same debates emerging in Britain a decade later.

On reflection, Clout argued that the experience of second home growth in France was 'yet another process' whereby the 'dispersed city' is encroaching on rural space. In the face of second home expansion, 'profound economic, social and visual changes have taken place in many parts of the French countryside and these may be expected to increase in magnitude in future years' (Clout, 1977, p.60). Overall, because the expansion of second home ownership is a component of wider social change, neither the lifestyles of rural nor urban populations can remain unaltered. Clout recognised that the second home phenomenon was not a passing phase in recreational fashion but rather a new and dynamic element of the housing market and perhaps a barometer for socio-economic change, not just in France, but across Europe and beyond. In this early work, he noted that a second home was often viewed as a sound capital investment, not only for the French, but also for the Dutch and the Germans on the Languedoc coast and for the British in the Dordogne and Perigord. Downing and Dower (1973, p.25) pointed out that by the early 1970s, 10,000 Britons owned second homes in Europe and further afield. Areas of growing British interest at this time included southern France, Spain, Portugal and Western Ireland. They argued that a rapidly declining rural population on the continent and the way in which foreign governments were keen to attract British tourists meant that:

> ... property overseas, both old and new, must be seen as a large potential source of supply of second homes for Britons, particularly at the upper end of the income/education scale (Downing and Dower, 1973, p.25).

In the 1960s and into the 1970s, British buyers remained minor players in the French housing market. However, in the 1980s this situation changed dramatically.

The work of Hoggart and Buller (1994a, 1994b and 1995) considers the move of British buyers into the French property market and their socio-economic impact. The number of rural properties purchased by Britons rose from 2,000 in 1987 to 14,000 in 1989 and some newspapers have speculated that as many as 200,000 Britons now own properties in France. Growing demand in the late 1980s was stimulated by the widening differential between house prices in the UK (particularly southern England) and the prices being paid for dilapidated rural properties in France. It was also apparent, according to Buller and Hoggart, that a desire to experience a more relaxed French life-style could be attributed to the books of Peter Mayle ('A Year in Provence', 1989 and 'Toujours Provence', 1990). The idealised rural image (and a perception of a loss of rurality in the UK) combined with positive feedback from existing owners to draw many Britons across the Channel. With reference to the earlier work of Clout, Buller and Hoggart point to the fact that although there is a high incidence of indigenous second home ownership in France, this is mainly confined to the coastal and mountain regions and to newer properties (some of which are purpose-built). Clout had earlier argued that inheritance played a major role in determining the distribution of second homes in France; Buller and Hoggart, however, argue that recent French generations have been increasingly cut off from rural origins. In effect, this change along with a continuing pattern of rural depopulation has left the rural property market open to foreign investment. In two recent papers, the authors have disseminated the findings of an interview-based study with 406 British buyers. The first of these (Hoggart and Buller, 1995) considers the impact of British owners on the French property market. In the second, Buller and Hoggart (1994b) examine the social impact of this encroachment.

In Britain, the status attached to rural living and the whole ideal of the rural dream has been questioned most recently by Cloke and Milbourne (1992). The lack of domestic market opportunities (particularly for isolated rural properties) and the erosion of Britain's rural myth (see Newby, 1980a) has meant that:

> It is inevitable that those who wish to acquire a rural haven will either have to compromise on their "dream" or look to satisfy their desires outside Britain (Hoggart and Buller, 1995, p.181).

Despite the large number of British purchases in France, market competition is limited. French second home buyers tend to acquire properties either through private negotiation (Dourlens and Vidal-Naquet, 1978) or inheritance (although this is declining) and whilst Britons have a preference for old, rustic property, the French are more likely to be in the market for new dwellings. In Normandy, for example, Hoggart and Buller were able to show that only 7.3 per cent of Britons purchased new homes. It should be noted that the work of these particular authors is not devoted entirely to the second home issue; almost half of the buyers interviewed claimed to be living in France *permanently*. However, this does not detract from the value of considering the impact of these acquisitions (many of which were for second home purposes). At the forefront of Hoggart and Buller's study is the argument that the demand from British buyers has been concentrated in areas of economic decline and the properties acquired were often dilapidated. In this context, new investment in these areas was welcomed, particularly as these properties had failed to attract domestic interest. Unlike the situation in Britain, buyers do not compete for mainstream housing and for this reason:

> Local estate agents are able to report that British buyers cause little friction in rural housing markets (Hoggart and Buller, 1995, p.189).

Instead, they have been instrumental in raising rural housing quality and releasing capital for local vendors who may subsequently buy a new home or improve their existing one. In general, British owners have created net quality gains in the rural housing stock although there have been some negative economic impacts. There seems to be little doubt that purchases in France have raised property prices. However, this is not the result of direct market competition, but because French estate agents specifically target British buyers in the hope of achieving higher prices. For this to have a serious impact on the property market, there would have to be competition for the same properties between British and French buyers. What has been observed is a *dual* property market where Britons are offered properties at higher prices than those offered to the French. This phenomenon has been quelled in recent years as a result of the British economic depression.

In effect, the demand for second homes amongst the British almost exists within an autonomous sub-market and Hoggart and Buller argue that the indigenous demand for second homes is likely to have a far more profound effect on the general property market. Similarly, since 1991 the number of Britons acquiring property in France has declined as a result of the economic recession (and the closing of the differential between British and French property prices) and the realisation amongst some buyers that the French rural idyll may not be quite as idyllic as Peter Mayle (1989, 1990) had suggested. But this is not the result of any pronounced hostility on the part of the French; the authors argue that:

> ... it has to be emphasised that the impact of in-migrant British purchases is quite different from that which is commonly associated with middle class inflows into British rural areas (Hoggart and Buller,1995, p.195).

In an earlier paper (Buller and Hoggart, 1994b), the way in which British property owners have 'integrated' into French rural communities was examined by the authors. The 'classical' perspective on this kind of social integration is that because the socio-economic make-up of in-migrants tends to differ from the host population, there is an inherent propensity for conflict and the disruption of social relations (Bonnain and Sautter, 1970; Forsythe, 1980). Newcomers often seek to protect their new home from change by either adopting a protectionist attitude (sometimes manifest as *nimbyism*) or by being critical of existing local practices (as noted by Albarre in southern Belgium). Buller and Hoggart argue that ensuing conflicts might be intensified in France because the British have a very different perception of rurality from the host population (Buller and Hoggart, 1994b, p.199). The ingredients for conflict were certainly present. The British tended to be younger than their French hosts and whilst the French were typically working class, the Britons were higher earning, middle class and petit bourgeois. Similarly, whilst the French saw the countryside as a working landscape, the newcomers tended to adhere to

77

the 'notion that rural space is primarily an aesthetic backdrop for residential location' (Buller and Hoggart, 1994b, p.201). To undermine the situation further, Buller and Hoggart demonstrated that British sensibilities often ran contrary to French rural values. For instance, Britons often registered complaints about the sale of horse-meat and hunting. The French complained that what the British failed to understand is the fact that these are deep-rooted traditions; in the case of hunting, the entire community becomes involved and it is an expression of social unity.

Despite these uncertain foundations, Buller and Hoggart were able to uncover very little evidence of any real social tension. In the communities, the British were often seen as just one component of a larger group of *étrangers* (strangers) and were certainly not as disliked as Parisians (with their easily recognisable cultivated accents and middle class or agnostic views). In fact, 94 per cent of British second home owners felt that they had been welcomed to rural France (Buller and Hoggart, 1994b, p.203). Remarkably, the greatest measure of resentment was observed *between* Britons. Some owners felt that the arrival of more Britons detracted from the French 'rural charm' whilst others found their compatriots distasteful, claiming that they treated the area like a colonial enclave and that this attitude was alienating locals. This was manifest in a 'compatriot fraternity' characterised by Britons who establish their own English-speaking networks and who do not learn French or mix with French people. However, it was clear that because of a lack of any clear social cleavage between British buyers:

> ... factors which separate those who seek English enclaves from those who attempt to integrate with their local community are linked more to value differences and to dissimilar objectives in buying a French home (Buller and Hoggart,1994b, p.205).

In conclusion, Buller and Hoggart argue that it is difficult for all but the most ardent francophiles to integrate into French rural communities. The linguistic and administrative barriers faced by many owners have proved too great to overcome and a number have now left France. Despite the friendliness of the French, they are a family-orientated people who may not have time for strangers. However, on a more positive note, it is clear that the motivations of British owners are not conducive to conflict generation; Britons rarely want to 'take over' (in the true nimby tradition) and their social relationships tend to transcend social divisions (something that is not achieved by Parisians and other French buyers). A key factor in promoting good social relations is the fact that there is a distinct lack of cross competition in the property market and the investment brought by the British is often welcomed. For these reasons, the authors argue that:

No matter what their goals and the limitations they face in achieving these, none of these British groups have socially disturbed their recipient French communities (Buller and Hoggart, 1994b, p.209).

As interesting and enlightening as the findings outlined by Buller and Hoggart are, it must be remembered that no direct comparison is possible with the situation in the UK. Direct competition for mainstream housing and the subsequent social disruption and anxiety contrasts markedly with the situation in France where the demands of Britons have been more easily accommodated and their class differences (and cultivated accents) have often been masked by language barriers and by a greater desire to 'fit in' (on the part of many owners). However, the point raised in the earlier paper (Buller and Hoggart, 1994b) that second home owners are not an homogenous group (even where there is little social differentiation) and many seek social integration in the host area may be applied equally to the domestic scene. It is clear in Wales, for example, that although some English in-migrants fail to recognise differing cultural values because of their Anglo-centric outlook, others are far more receptive to the cultural and linguistic values of the host population. Analyses of the socio-economic make up of settlers or second home owners should not seek to tar all these groups with the same brush.

Greece

It is probably not surprising that little academic attention has been focused on the second home situation in the EU's most easterly member state. In 1988, 12 per cent of households owned a second home in Greece; this figure may be viewed as particularly low given Greece's track record in land use planning and Government-promoted moves to extend the social scope of ownership in the 1960s. However, the growth of second home ownership is at least partially rooted in economic prosperity. Robbins (1930) has illustrated that increases in the amount of time spent engaged in leisure activities is a function of higher incomes (Robbins, 1930, pp.123-129) and it is only relatively recently that the amount of leisure time (and real income) has begun to rise in Greece, a country which still had the lowest gross domestic product per capita in the EU in 1992 (Eurostat, 1994, p.11).

In 1980, Psyhogios undertook an empirical study of the second home phenomenon in Nea Makri, an attractive coastal settlement just 33 kilometres from the centre of Athens. A principal factor in the expansion of second home ownership in Greece (and particularly around Athens) was the failure to allow for the recreational needs of urban populations during periods of rapid urban growth during the early years of the twentieth century (Psyhogios, 1980, p.86). More recent rises in real incomes meant an increased demand for leisure activities; this demand could not be accommodated in Athens and so wealthier urbanites began to look further afield. The vast majority of second homes in Greece are concentrated around Athens and Thessalouiki and are mainly coastal (Psyhogios, 1980, p.88). The Census of 1971 revealed that there were between 35,000 and 40,000 second homes in the greater Athens area and of these, 88 per cent were on the coast whilst the remainder were in the uplands. The distribution was driven largely by consumer taste with certain small fishing villages being transformed, by a process of gentrification, into fashionable resorts. These second home resorts are characterised by dwellings ranging from single family houses to multi-rise buildings and by high residential densities.

A serious Greek problem has been the lack of regulatory standards and planning control (Psyhogios, 1980, p.94). Second homes have tended to be *purpose-built* and their development is not as closely related to rural depopulation as it is in many other European countries. The lack of planning control is rooted in the way in which successive Greek governments have pro-actively encouraged new development and the social democratisation of second home ownership. In 1967, for example, a Royal Decree stated that *lyomeno* dwellings (demountable vacation houses) could be erected anywhere not covered by a 'master plan'. This move was intended to extend the opportunity of owning a second home to middle and low income families, but instead, it generated a rush in land speculation on the part of developers who subsequently erected more expensive (and profitable) dwellings for higher income groups. Other types of second home development in Greece include multi-rise flats (which have emerged more recently in response to increasing coastal property prices) and villas which attract higher income urbanites and are located in the older, more traditional, vacation areas (Psyhogios, 1980, p.97). The pro-development ethos of Greek planning was reflected in Department of Housing Circular 22 (24/4/77) which listed those types of areas suitable for new second home development. The guidelines simply stated that second homes should be sited in areas of maximum recreation potential (that is, with suitable climates and access to water), should be near existing settlements and easily connected to road networks and should avoid coasts with polluted waters. Clearly, these guidelines were geared to the recreational needs of the users and gave little credence to wider socio-economic factors. However, Circular 22 did state that new development should be steered away from forested and rural areas, not because of any concern for the rural population, but because of the difficulties in providing infrastructure support in these areas. Some 'rural' agglomerations had been conceived in micro-towns (Psyhogios, 1980, p.104) which appeared to meet with approval from planners, the conservation lobby and some weekenders. In addition to Circular 22, Decree 947 (26/7/79) stated that designations of these 'suitable' areas could be made in a number of ways. The Ministry of Public Works could make an immediate designation, whilst designations could be proposed by local authorities or made under private application. In effect, it was possible to build second homes almost anywhere, providing they were suited to the needs of the users.

In the study of Nea Makri, Psyhogios claimed that the growth in second home ownership had brought a range of benefits and very few problems. The creation of wealth for local people (through lucrative property sales) had meant that they were advantaged rather than disadvantaged in the property market. There was certainly 'no evidence for any responsibility [on the part] of second home development for local depopulation in the area' (Psyhogios, 1980, p.143) and quite the opposite situation seemed to have developed, with growth in the permanent population running parallel to the growth in second homes. Social tensions were practically non-existent, but this is perhaps not surprising given the economic benefits brought by incomers and the lack of socio-cultural differences between groups. Nea Makri is just 33 kilometres from the centre of Athens; other European studies have examined groups from entirely different regions or even countries. The situation in Nea Makri does not fit easily into the more general cost/benefit models of second home ownership and Psyhogios's interpretation is intended to reiterate the desirability of new development. Given Greece's trailing economic position in the EU, the continuing dominance of the pro-development lobby is to be expected. However, the importance of careful planning for second home demand is still recognised and Psyhogios endorses the argument by Martin (1972) that only by:

> ... anticipating and by planning for the demand, can the inflow be controlled. If matters are left as they are - to the interplay of market forces - areas will have a settlement imposed on them, dictated solely by wealthy urban interests (Martin, 1972).

In Greece, however, it has been difficult to plan for demand in a co-ordinated way. The main reason for this is the lack of any local input into the planning system. Planning control is entirely in the hands of the national government which has tended to give insufficient attention to the siting of new development. In the 1970s, for example, second homes were zoned in the same area as a proposed international airport and it is a failure of the Greek planning system that all development is deemed beneficial irrespective of siting and local concerns. The belief that second homes have a 'peculiar strength as a potential contributor to the social and economic development of the receiving areas' (Psyhogios, 1980, p.158) has underpinned government attitudes towards this type of development. In the 1980s, it was argued that the most significant factor advancing the Greek economy was the 'development of private property (land and housing), especially in areas of second home and tourist demand' (Getimis and Kafkalas, 1992, p.80). However, irrespective of the costs and benefits of second homes, Psyhogios argues that all systems of housing priorities should consider 'the supply of a satisfactory first home before the provision of a second' and planning policies should not 'ignore the majority of people for whom second home ownership is but a dream' (Psyhogios, 1980, p.161).

Spain

It has been claimed that the growth in second homes represented one of the most significant changes in the Spanish housing market in the 1970s (Barke and France, 1988; Morris, 1985). There were estimated to be 2 million second homes in Spain by 1981 and even in the early 1970s, 17 per cent of Spaniards had access to a second home leading some observers to comment that 'ownership is almost as much a part of Spanish life as it is that of the French'. As in France, however, many Spanish second homes are owned or used by Foreigners. In the 1970s, Spain was experiencing a period of rapid expansion in 'residential tourism' that was largely unaffected by amendments to planning law (particularly, changes to the *Ley de Regimen del Suelo* 1956) which had no real impact until the 1980s. In effect, the 1970s bore witness to Spain's last phase of largely uncontrolled development.

Although the Spanish population census (like others across Europe) has clear flaws where the enumeration of second homes is concerned, it can be estimated that between 1970 and 1981, the number of second homes in Spain rose by 147 per cent from 796,000 to almost 2 million. The 1970s saw a diffusion of domestic-market second homes from larger, more accessible settlements to the smaller, more remote ones (Barke, 1991, p.14). At the same time, new purpose-built (and often foreign-owned) second homes were increasingly being concentrated in the larger settlements. There was a clear distinction between foreign owned second homes and more recent acquisitions made by Spanish nationals; this particular phenomenon and the notion of a dual market was examined by Barke and France (1988) in the Balearic Islands. In terms of the up-turn in the domestic market in the 1970s, two processes were at work. First, *urban* dwellers, experiencing greater economic prosperity since the 1960s, were purchasing inexpensive 'hobby farm' plots in the interior, which often had sheds and farmbuildings which could gradually (in an *ad hoc* fashion, avoiding planning control) be converted into inhabitable dwellings. The second process involved *rural* families moving to urban dwellings but retaining part of their land for recreational use (Barke, 1991, p.14). According to Barke and France (1988), 'it is ironic that such a trend should be under way at a time when urbanisation has been increasing' (Barke and France, 1988, p.144) and large numbers of Spaniards have their principal residences in urban apartment blocks. A predominant concern in Spain has been the growth and distribution of second homes (examinations of impacts have been confined to those areas experiencing extraordinary rates of growth; see Jurdao Arrones, 1979) and the importance of landholding in relation to these aspects of the phenomenon. Where landholding patterns involve the ownership of land in fewer hands, development tends to have been more extensive. This has been observed, for example, in the centre north and the west of the country which have experienced the highest rates of growth. However, in terms of the total number of second homes, the area of highest concentration extends 'from west of Madrid through to the Mediterranean coast in the east and including the Balearics' (Barke, 1991, p.15). Provinces adjacent to Madrid have peaks reaching 35 per cent of the housing stock with peaks of just over 30 per cent around Barcelona. These concentrations reflect the presence of foreign buyers. In the north of the country, and particularly in the Basque region, there are far fewer second homes, partly because of the perceived strength of nationalist and separatist feelings. Patterns of ownership are clearly related to the significance of regional tourism; however, the second home phenomenon diffused widely in Spain in the 1970s and now relates largely to the redistribution of the Spanish population (Barke, 1991, p.17).

An important feature of second home growth in Spain has been the dominance of certain coastal regions; a pattern sustained by foreign investment. In Malaga province, for example, the proportion of second homes grew from a base of 8.8 per cent in 1970 to 18.6 per cent in 1981. The cause/effect relationships often seem clear-cut in Spain with apparently straightforward correlations between economic decline and second home growth or the attraction of coastal regions and foreign acquisitions. However, on occasions, conventional wisdom appears to break down. Away from the coast in Malaga province, although rural depopulation has clearly taken place, few dwellings have been bought as hobby farms and even fewer have been retained by families moving into the urban areas (Barke, 1991, p.18). It appears, therefore, that replacement is not always the natural successor to abandonment and where the foreign tourist base is overwhelmingly dominant, even indigenous buyers may gravitate towards it, attracted by recreational opportunities and the vastly superior service infrastructure. In this instance, patterns of acquisition fall outside the classical interpretation and buyers are willing to sacrifice the opportunity of acquiring cheap rural property if the attractions of another, far more expensive and urban-based option, are that much greater. It might be desirable to compare this scenario with the situation here in Britain and consider whether it might be possible to control the demand for rural second homes by increasing the recreational facilities available in exporting regions. However, such a comparison is largely invalidated by the differing socio-cultural contexts and the lack of any appreciation of the value assigned to rural space by various Spanish social groups. It is interesting to note though, that Psyhogios (1980) attributed much of the growth in second home demand in Greece to the failure of planning authorities to allow for the recreational needs of urban populations.

In Malaga, second home distributions are associated with new-build rather than abandonment and replacement (Barke, 1991, p.19) and therefore land use planning has played a far more dominant role in shaping development than in many other Spanish provinces. Elsewhere, it is still clear that the second home market is characterised by a dualism whereby foreigners dominate the larger (often coastal) settlements and Spanish ownership is more characteristic of smaller, often remote settlements. These patterns have been underpinned by rural depopulation in the interior and burgeoning residential tourism on the coast which have generated a division between casual, individual use and large scale, highly capitalised development (Barke, 1991, p.20). The literature examining the second home phenomenon in Spain has been fairly limited in scope and considerably 'underestimates the complexity of the phenomenon and the variety of conflicts resulting from its rapid growth in Spain in the 1970s'. An area of growing concern has been the:

... social and cultural dislocation experienced by local communities where a large influx of temporarily resident strangers takes place, especially if they are foreigners (Barke, 1991, p.14).

To date, little work has been carried out on this aspect of second home growth in Spain, although some clues as to the socio-economic impact of foreign investment might be gleaned from the work of Hoggart and Buller in rural France. However, much of the foreign investment in Spain has been coastal rather than rural and it might be speculated that the types of British visitors spending their holidays in rural France differ markedly from those visiting the Spanish coast. Barke (1991) notes that the full impact of the second home phenomenon has not yet been fully assessed in Spain and it remains to be seen what impact the planning and environmental policies adopted in the 1980s and implemented in the 1990s will have on second home development and in meeting the objections of an increasingly vocal indigenous population' (Barke, 1991, p.21).

Sweden

Bielckus (1977) argued that regulatory measures to control the growth of second homes across Scandinavia became increasingly important in the 1960s in order to safeguard the countryside whilst continuing to accommodate demand. The changing situation in Sweden received considerably more attention than in neighbouring Denmark or Norway. By 1970, there were almost half a million second homes in Sweden (Bielckus, 1977, p.35), owned by 22 per cent of Swedish households (Downing and Dower, 1973, p.20).

The distribution of second homes in Sweden was first mapped in 1938 (Ljungdahl, 1938) and in some respects, still reflected the semi-nomadic tradition of Swedish farmers, moving herds from winter to summer pastures (Bielckus, et al, 1972, p.132). More recent studies (Larsson, 1969) have emphasised the division between urban exporting regions and rural importing regions, showing that two-thirds of all Swedish second homes are owned by people living in blocks of flats. In the 1960s, rural depopulation and the movement away from the land freed 10,000 farms each year and 30 farm houses became available each day. This only partially satisfied a growing Swedish demand and between 1967 and 1970, a further 55,000 plots of land were made available for second home use (Bielckus, 1977, p.38). A particular oddity of the Swedish situation is the concern over the style of second home dwellings. The red-roofed crofter cottages of Dalarna were so popular in the early 1970s, that many were dismantled and moved nearer urban centres. In the 1960s, observers were already noting that the supply of rural dwellings was not inexhaustible and for this reason, the second home aspirations of individuals and the needs of rural communities were not always compatible. As in Denmark (see previous discussion), there were growing concerns that second home ownership and the privatisation of rural land was creating a potential problem in terms of public access. In 1962, the Swedish Government appointed a review commission with a view to establishing an accurate picture of the distribution of second homes and a basis on which to establish coherent planning policies (Williams, 1974, p.15). The subsequent report (Statens Offentliga Utredningar, 1964) revealed that 85 per cent of second homes were located in the central and southern parts of the country. In 1968, the National Swedish Land Survey Board estimated that there were 420,000 second homes with the principal concentrations around urban centres; in fact, one-third were within 20 kilometres of a settlement with more than 25,000 persons. However, this situation was beginning to change. In 1968, Aldskogius carried out a study of second homes paying particular attention to the types, age and rateable values of buildings and travel distances from first to second homes. On a regional level, he was able to demonstrate that *new* developments were dominated by owners from outside the importing region, indicating that people were beginning to travel greater distances to their second homes (previously, 65 per cent of Swedes had lived within 50 kilometres of their second home).

An academic concern with second homes in Sweden was juxtaposed with an official concern for land-use planning. The earliest aim of this planning in the 1960s was to 'achieve a balance between vacation housing and provision for active outdoor pursuits' (Bielckus, 1977, p.42) with public access taking precedence over the development of second homes. Pressure on the coastline was seen as a particular problem in this context. The release of land for second home development was also occurring in a piece-meal fashion which worked contrary to the objective of long-term effective planning. For example, the extreme fragmentation of land-holdings around Lake Siljan in Dalarna (a particular historical phenomenon) was hindering the purchase of appropriate plots for second homes and causing demand pressure on the existing dwellings which

86

became much-coveted (Bielckus, 1977, p.43). These types of concerns generated pressure for legislative change and the control of real estate acquisition (see below). However, the first moves to control the development of second homes occurred at the level of regional planning. The earlier Government research commissioned in 1962 had shown that a large number of second homes were located in close proximity to urban areas and with the expansion of these areas, there was evidence of some incorporation of second homes into primary residential areas. Around Stockholm, former second homes (many of which lacked proper facilities for waste disposal) were reverting to first home use, providing poor quality suburban accommodation for lower income groups. In order to prevent this from happening in future years, the 1966 Regional Plan for the Stockholm area specified that by the year 2000, no second home developments would be located within one hour's drive of the city. By 1968, Aldskogius had already noted that the tendency to acquire second homes in more remote areas was increasing and therefore, the problems associated with this particular change of use were likely to diminish as a result of improving transport infrastructure and regulatory controls.

Other planning measures were also implemented at this time including the regulation of new development in the open countryside. It was recognised that many of the problems associated with second homes in the 1960s (particularly on the Swedish coast) were the legacy of short-sighted planning decisions made in the 1930s. When the holiday settlement of Herrviksnäs was completed in 1935, for example, its siting prevented general access to the coast; in contrast, the development of Hanskroka (1956) posed no such problems and the settlement was far less unit-dense (Bielckus, et al, 1972, p.133). Development in the 1960s and 1970s was rarely permitted within 300 metres of the shoreline. In the uplands, although dense natural woodland could be used to conceal new development, it was clear that it should not be used to conceal *poor quality* development. Increasingly, planning control was used to maintain standards with plans being rigorously tested before buildings were erected; far greater emphasis was placed on the provision of adequate facilities for water supply and waste disposal. By the mid 1970s, typical second home developments comprised between 100 and 200 dwellings (the 200 limit was rarely exceeded). In 1974, Williams argued that this move towards developing second homes in purpose-built 'communities' should be examined with reference to its 'relevance in the British situation' (Williams, 1974, p.35). By the end of the decade, it was clear that demand for second homes was levelling off. Improved camp-site facilities across the country meant that an increasing number of tourists were spending their vacations camping in the mountains; across Scandinavia (and particularly in Sweden and Norway), vacation chalets for hire were also becoming more popular, providing a cheaper alternative to purchased second homes (Bielckus, 1977, p.44).

On the whole, the Swedish government became far more concerned with the way second home developments were planned and regulated from the 1960s onwards. By the 1970s, far more attention was being paid to infrastructure provision and the striking of a careful balance between conservation and incorporating second homes into the countryside. Increasingly, 'maintenance of the amenity value' became the main concern in the planning system; second homes were an integral part of the tourist industry, but could not be allowed to disrupt recreational opportunities for other tourist groups, either by damaging the environment or by limiting public access to the most attractive areas. Bielckus neatly summarised the situation that had been reached by the mid 1970s:

> In Scandinavia second homes have become a sufficiently widespread phenomena to warrant specific legislation. Particularly in Sweden, the favourable population-land ratio and attempts at an equitable distribution of economic resources create an environment in which the desirability of second homes as a form of land use need not be questioned at this time (Bielckus, 1977, p.44).

However, another parallel trend had been developing in Sweden for a number of years; this was a trend towards social as well as natural conservation, which emphasised the need for compromise between individuals and society at large. Where this cannot be achieved through planning control, restrictions on the acquisition of real estate for certain purposes are applied. A growing problem in Sweden in recent years has been the demand for second homes from non-Swedish nationals. Where this has created particular problems (such as upward pressure on house-prices) and the demand for 'secondary residences' is high, permits (required under the *Law on the Acquisition of Real Estate* 1992) for acquiring second homes will not be granted to foreigners. To offer additional protection to some communities, the *Law of Option to Purchase Real Estate* 1967 allows communities to purchase properties which come on the market in areas where the demand for second homes is high and where there is a shortage of primary residences. Under this legislation, communes must declare their intention to buy; an intention which can be challenged by the vendor. Where disputes arise, Central Government acts as arbitrator and it is not uncommon for appeal decisions to go in favour of the vendor. Critically, these measures have been shown to be insufficient in many parts of Sweden. On the west coast, for example, heavy demand for second homes (in the mainstream housing stock) is causing the elevation of house prices, leading to an inability, on the part of younger households, to compete in the housing market. For locals in general, higher property values mean burgeoning taxes as their properties are re-assessed in higher tax-bands (the *Law of Assessment of Real Estate* 1979 does not assess individual dwellings and general taxation is based on an aggregate of market values across an area; therefore increasing tax costs are shouldered by all owners, temporary or permanent). In this situation, it is clear that no compromise has been achieved between the aspirations of individuals and the needs of the local communities.

In Sweden, new second home developments have been increasingly well-planned and, arguably, the objective of achieving a balance between conservation and recreation has been met (Jacobs (1972) recognised that there were lessons to be learnt from the Swedish experience). However, the problems brought by competition for existing housing in some areas have been less easy to regulate despite successive moves to control the acquisition of real estate. Legislation designed at the national level may not always have the desired effect at the local level (this has been apparent on the west coast). This final point acts as a reminder that national instruments must be grounded in a thorough understanding of local circumstances.

Norway

Far less is known about the second home situation in Norway than in Sweden or Denmark. In 1970, there were estimated to be 170,000 second homes in Norway (17 per cent of households), although Bielckus (1977) concedes that data regarding second homes in Norway is undermined by the problematic way in which tourist accommodation is classified. Accommodation listed as *hytter* (or 'huts') may range from a simple private cabin to a small hotel (Bielckus, 1977, p.37).

Second homes in Norway are often an expression of the desire on the part of Northern people to enjoy light summer months in the south of the country. Therefore a broad pattern of distribution can be identified; however, in the importing region in southern Norway, development tends to be scattered and has taken place with little planning or control in the past.

It is only in recent years that Norwegian planning authorities have experimented with planning controls. In Hallingdål, for instance, a two year ban was imposed on hut building in the 1970s whilst the local authority investigated the ways in which new development might be clustered (drawing on experience in Sweden). The definitional problems point to the fact that in Norway, second homes are seen in the broadest context of recreation; they are sometimes examined at the local level (in planning or economic reports) but rarely at the national level. There is little recent research (English) regarding the second home phenomenon in Norway, although Ouren (1969) estimated that an annual growth rate of 10,000 units was likely to be sustained in the 1970s.

Other European States

In some prior second home studies, additional references have been made to the second home phenomenon in other European countries, although very little detail is provided. In Czechoslovakia, for example, Gardavský (1977) points out that city dwellers had attempted to establish second homes since the turn of the century, creating recreational hinterlands in a number of city-regions. The basic lack of land use planning in both the Slovac and Czech Republics became the focus of considerable concern by the mid 1970s (Gardavský, 1977, p.72). The extrapolation of growth trends suggested that a number of new (but undeveloped) 'recreational areas' would be created by the mid-1980s, offering the opportunity to steer and regulate future development. This was made possible by a growing pool of public funds with which to pay for improved recreational planning. However, it should be noted that Gardavský's suggestion that it would be desirable for *existing* cottages in areas of special attraction to be converted to second home use rather than adding non-vernacular, out-of-context, developments was certainly out of step with the dominant western view. The social problems that this might have caused may well have been ironed over by assertive authoritarian control of the type which led to the forced relocation of some rural populations in the former communist states. However, a greater exchange of ideas today may help avoid the particular socio-economic problems associated with patterns of rural change across Europe, both East and West.

In the 1970s, a property licensing system operated in the Netherlands (Pyne, 1973, 44) along similar lines to those operated in the Channel Islands and in some parts of Scandinavia. However, there is no recent research examining the scheme today or legislative amendments which might have been implemented given various EU directives regarding the freedom of movement (and residence) between member states. Bielckus *et al* (1972) have noted that the Dutch experience of the second home phenomenon has been similar to that of the UK with a number of noticeable differences (Bielckus, *et al*, 1972, p.136). A large proportion of the Dutch population lived (and continue to live) in urban areas, whilst there were few agricultural areas in which farms or other rural properties were becoming vacant. However, Rogers (1971) observed that some second homes were becoming available in the reclaimed *polders* where too many new villages had been planned in response to over-estimates of the future employment capacity of agriculture (Rogers, 1971, p.277). It is inevitable that surplus rural dwellings, in the Netherlands or indeed in the UK, may eventually be used for a purpose that was not originally intended or envisaged. It should be noted that in some circumstances it may not be desirable to restrict this change of use if it means that dwellings remain empty and fall into disrepair.

Particular studies of second homes in Austria and around the Austrian-German border have been undertaken in recent years (Bennett, 1985; Salletmaier, 1993) although these have concentrated on the macro-economics and the more theoretical demarcation of 'recreational space'. The activity of German second home seekers has attracted less attention in Germany than in other European states. The importance of German buyers in southern France was noted by Clout (1969, 1977) whilst their impact in Sweden has attracted more recent attention. Increasingly, movement between European member states means that it is becoming more difficult to place the second home phenomenon within rigid national boundaries. It is still apparent, however, that the choice of policy response is still very much in the hands of national legislators, reflecting the structure of Government and the dominant political ideology that prevails.

Despite the contrasting socio-economic and cultural contexts in which different second home experiences have evolved, it is still useful to list the most salient points, some of which may have some relevence in England, Scotland and Wales (and across Europe generally). Albarre's study of second homes in Belgium demonstrated that the problems posed by social change (of which second homes are a component) are often balanced by the new opportunities offered. In his particular case-study, new opportunities for conservation were generated by the arrival of newcomers. It is the purpose of planning to ensure that such opportunities are maximised and therefore the scales are tipped in favour of rural communities. In

Denmark, increasing recreational opportunities is seen as a legitimate planning goal and second homes must be seen in this context. However, second home demand can only be accommodated after wider social objectives have been achieved. Change of use legislation in Denmark attempts to ensure that this is the case. The French experience of second homes was one of purpose-built resorts and acquisition of properties through inheritance until, that is, the British arrived and began purchasing old derelict rural property, and reminded everyone of the apparent socio-economic benefits that this type of investment can bring. A similar pattern of use change in surplus rural dwellings occurred in the Dutch polders in the 1970s. It is clear that where change of use is controlled (or might be controlled in the future), it should be done sensitively. There would be little point in barring the change of use of dwellings that will otherwise remain empty and fall into disrepair. The French situation also acts as a reminder that second home owners are not an homogenous group. The different motivations of second home seekers dictates the social contribution that they might make in host communities; but how might planners allow for these differences?

In Greece, it was clear that if second home villages are a desirable option in some areas, they need to be carefully planned. Arguably, however, British planners might learn more from experiences in Sweden than Greece in this respect. One astute point made by Psyhogios however, was the recognition that strategies for second homes must invariably take a back-seat to the development of primary residences. In Spain, the dual nature of the second home market (as in France), provides further evidence that residential tourism and the ownership of second homes is a growing international phenomenon. Hoggart and Buller point out that when domestic second home aspirations cannot be realised, demand may drift abroad. From a national perspective, this may have an adverse impact on the balance of payments; from a European and perhaps more altruistic perspective, it may not be seen as desirable to plan in such a way at home as to export domestic problems to other member states. The case in hand is certainly that of German buyers in Sweden. In Sweden, the now 'classical' problems (of house price inflation and social disruption) remain and are accentuated by a tax system which appears to 'charge' permanent residents for the number of second homes which they are forced to endure. It appears that real estate acquisition restrictions have failed to close the floodgates on either domestic or foreign purchasers. More success had earlier been achieved in negating the adverse effects of second homes through land use planning and as far as the real estate restrictions are concerned, they appear to represent a policy approach conceived at the national level that does not always operate smoothly at the local level. New planning measures, either in the UK or abroad, could of course suffer the same fate. It would be important in designing any new legislation to ensure a synthesis of both national and local perspectives in order to create a policy framework which is both sensitive and equitable. However, this would be an extremely ambitious goal.

Finally, in Finland the case of Åland demonstrated how measures (in this case, based on *property rights*) may be implemented that reflect the dominance of the local rather than the national perspective. These, of course, may only be attainable in countries where relatively devolved government structures exist. All policy responses operate in different legislative contexts. In Scandinavia (and particularly in Finland), devolved regional structures emphasise local perspectives and offer more opportunities for prioritising local interests. On the other hand, where centralised government and a dominant free-market ideology prevail, the national perspective is emphasised.

Although there are clear benefits to be derived from an exchange of ideas, differing socio-economic and political circumstances mean that individual countries must formulate their own tailor-made responses to the problems they face (Allen, *et al*, 1999). Some of the current responses examined in this Chapter are summarised in Figure 2 overleaf.

Figure 2: Comparison of European issues and responses

County	Principal concerns	Key issues	Policy responses
Markets dominated by domestic buyers			
England and Wales	Social/economic/regional	A competitive rural housing market characterised by higher income second home seekers and demand focused in the mainstream housing stock	The provision of non-market housing and occupancy controls employed by local authorities
Markets dominated by domestic and foreign buyers			
Denmark	Recreational/social/environmental	Planning for second home development and recreation; the issue of foreign purchasers accentuating housing shortages	The planning system distinguishes between residential property types and use change controls may be used to protect local interests
Finland	Social/regional	Foreign purchasers accentuating housing shortages (especially in some provinces)	A devolved system of government makes it possible to define local property rights
France	Social/economic	Domestic demand for rural housing following the trend set by the British in the early 1990s	*No information*; but the French planning system has a far more positive role in promoting development than its British counterpart
Spain	Economic/environmental	Large rises in second home demand and the development of a dual market, with a large number of foreigners purchasing purpose-built second homes	*No information*
Sweden	Recreational/social/environmental/economic	Planning for second home development and recreation; the issue of foreign purchasers (particularly German nationals) accentuating housing shortages	Evolution of a planning system geared to second home development and more recent controls over the acquisition of property, particularly second homes by foreigners
No market information			
Norway	Unknown	Development of good-practice planning	*No information*
Greece	Economic	A concern for the way in which second home developments are planned and sited	Planning guidance issued in the late 1970s
Belgium	Social/unknown	A competitive rural housing market characterised by higher income second home seekers	*No information*

7 Housing Policy Responses in the UK

The final point made in the last Chapter of Part I (which has provided a defining theme throughout the discussion so far) should demonstrate that responses to second home growth need to be based on a dual approach in which certain strands of development are encouraged whilst others are actively discouraged. Throughout the second home literature, this distinction between potential responses has been continually emphasised. In the last Chapter, we considered how other European countries (outside the UK) were addressing the second homes issue. The purpose of this Chapter is not to consider how effective policy might be formulated within the UK, but rather to examine the suggestions put forward in the prior studies, many of which have been pre-empted by more recent developments in housing and planning policy. Chapters Seven and Eight examine in detail the way in which today's housing and planning policy framework could best maximise the positive impacts of second homes whilst minimising the economic, environmental and social costs; the perspective is made predominantly from England and Wales. More generally, options for controlling the growth in second home ownership fall into three broad categories; economic and social development, housing policy options and planning and development control (Dower, 1977, p.161).

Economic and Social Development

Clearly, although the growth in second home ownership might be viewed as a complicating factor in processes of socio-economic change, it may still be argued that the inability of locals to compete in the housing market (against Shucksmith's 'more prosperous groups') is caused by the weak state of the local economy, not by second homes, and therefore emphasis should be placed on boosting the local economy rather than using second homes as a 'convenient scape-goat' (Dower, 1977, p.161). In any economic strategy, second homes will have an important role to play as part of the wider tourist industry. One policy option, therefore, might be to promote second home ownership in such a way as to maximise economic benefits (whilst diverting demand away from existing housing stock). This move would not be without its critics. In the 1960s and 1970s, nationalists in Wales were keen to promote the myth that 'second homes and the whole tourist industry were an unmitigated disaster for the Welsh' (Coppock, 1977, p.200). Clearly, analysis of the effect wrought by second homes outside the effective housing stock demonstrates that this is not the case and whilst there is a good economic argument for promoting development in some areas, it is also generally accepted that second homes do fulfil a real need (Coppock, 1977, p.211). Downing and Dower have argued that second homes satisfy recreational benefits and pose the following question to those opponents who argue for prohibitive policy measures:

> If desire for second homes is frustrated, will some other effective outlet for recreational impulses be available: and will its implications be more acceptable on balance than those of second homes? Or if impulses remain frustrated, who will justify the human and societal loss? (Downing and Dower, 1973, p.29).

For this reason and because of the underlying economic logic of promoting second home ownership, control needs to be balanced with a means of satisfying demand. Jenkin (1985) follows this line and argues that a 'policy is needed which attempts to absorb the demand for second homes while at the same time directing that demand away from those communities which are at risk from the intrusions' (Jenkin, 1985, p.70). This would be achieved by identifying sites suitable for purpose-built second home developments and using development control to discourage developments in unsuitable areas; control of the layout and appearance of sites would neutralise the environmental impacts (Jenkin, 1985, p.74).

The success of this strategy, in maximising economic benefits and minimising social costs (by diverting demand away from mainstream housing stock) would hinge on two points; first, the attractiveness of purpose built second home 'villages' and second, the willingness on the part of elected assemblies or policy executives to assign specific protection for existing housing (Pyne, 1973, p.52). On the first issue, Davies and O'Farrell have shown that second home seekers will be attracted to developments of fairly high residential density (in the form of 'holiday villages') if these developments are correctly sited (Davies and O'Farrell, 1981, p.108). In their particular analysis, this meant close proximity to the coastline but it would probably also involve proximity to a range of other facilities such as settlements or landscapes of specific interest, inland water surfaces, key tourist attractions and services. Increasing supply in this way might attract people away from static caravans and have a positive environmental impact (Pyne, 1973, p.51).

As Pyne (1973) has noted, diverting demand in the way described above will only be effective if added protection is given to existing housing which is likely to draw the attention of potential second home buyers. The prior studies tend to concede that this would involve change of residential use being brought under planning control; this issue is introduced later in the book. It is clear however that promoting the tourist industry (of which second homes are a part), provides a significant part of a longer term strategy which must be to raise local income levels (Shucksmith, 1983, p.187). Dart's 1977 report in Scotland argued that local government should indicate clear support for tourism in local plans, provide tourist infrastructure (such as roads, harbours, ski slopes and water sports areas) and have a direct hand in second home development and the promotion of the area as a tourist destination (Dart, 1977, p.77). In their earlier report, Downing and Dower (1973) even advocated direct public provision of second homes, a move that seems somewhat unlikely today (Downing and Dower, 1973, p.36). Encouraging new employers in the area (in all sectors, with the service sector potentially being the most lucrative in many rural areas) will improve employment conditions with increasing demand for labour and higher wage levels. A stronger local economy would generate an upward pressure on both house prices and the supply of new housing and place locals in a far better competitive position in the housing market. All these changes are of course desirable but may not bring a panacea for various socio-economic ailments in the short term. It is clear however, that over-restriction of second home development would run against the tide in terms of sound economic planning which at the end of the day, should seek to maximise rather minimise potential benefits. It is also true however, that:

> The basic test of the desirability of second homes must be the local interests and the consideration of rural amenities ... the positive promotion of holiday villages as a form of tourism will not [on their own] create the economic salvation of rural areas (Williams, 1977, pp.60-63).

Housing Policy Options

The *housing* policy options suggested in prior second home studies tend to be concerned with the provision of housing alternatives (for local residents) in the face of second home demand and general housing shortages or, ways in which housing (or related) policy might be used to curb or even halt the demand for second homes. Because some locals fall into Shucksmith's (1990b) 'low income, low wealth' consumption class, they may not be able to successfully compete for market housing against more prosperous incomers; it follows that one solution is to provide non-market alternatives.

Providing Non-Market Housing

Non-market (and usually rented) housing has traditionally been provided by local authorities and the strategy of building rural council housing was generally successful in sustaining local communities. However, this situation has now changed and whilst the stock of local authority dwellings has been reduced (typically by a third) by the right-to-buy since October 1980, similar reductions in local authority finance (and the inability of local authorities to invest capital receipts from sales in new development programmes) has meant that the traditional rural role of local authorities in direct provision has become little more than an historical footnote. Local authorities have, in recent years, been asked to play a new housing role; that of development 'enabler' (Goodlad, 1993; Bramley, 1993). Since the creation of the Housing Corporation in 1964 and the establishment of a grant framework in 1974, voluntary sector housing associations have come to play an increasingly important role in the provision of 'social' (i.e., non-market and 'affordable') housing. By the mid-1980s, these associations were actively engaged in new rural housing schemes across Britain. However, the favourable financial regime brought about by the introduction of the Housing Association Grant (HAG) in 1974 was brought to an abrupt end in the Housing Act 1988. From the following financial year, full public subsidy was replaced by a form of public 'deficit funding' whereby an ever larger part of the cost of new provision would have to be met by the private sector. Over the last ten years, grant rates have been incrementally reduced and these reductions have threatened the affordability of new schemes. In this context, local authorities and associations have been encouraged to 'work together' (see Fraser, 1990) in order to reduce the cost of provision, sustain supply and ensure that new housing schemes remain affordable. This strategy may involve the transfer of public land or loan sanction from the local authorities to the partner associations. It may also involve the local authority developing 'local housing need strategies' (planning-based, which are written into local plans) which generate additional sources of affordable housing on private development sites (i.e., achieving site-specific quotas through planning negotiation) or on rural 'exception' sites (i.e., off-plan; Joseph Rowntree Foundation, 1994; Barlow and Chambers, 1992; Barlow, et al, 1994) which is subsequently developed and managed by registered housing associations (whose long-term social objectives ensure that housing serves community needs in perpetuity).

However, these partnership strategies have proved problematic, particularly in rural areas where there have been difficulties in securing the necessary input of private finance on exception sites due to the inclusion of Section 106 agreements (RDC, 1995) and where poor development plan coverage has worked against the negotiated strategy (Gallent, 1997). In can be argued that the synthesis of housing and planning policy (Tewdwr-Jones, et al, 1998) generates an inadequate supply of additional rural housing and therefore cannot accommodate all those local residents excluded from the private property market as a result of increased local competition. On the other hand, the contribution of housing associations in providing homes for many rural families should not be underestimated and in this context, the Conservative government's purchase grant schemes for association tenants, introduced in the Housing Act 1996, may prove to be all the more damaging in some rural areas, particularly as it may discourage some landowners from releasing land for housing on some off-plan sites and, like its predecessor in 1980, it may substantially reduce the supply of non-market housing in the countryside.

It is these types of policy moves over the last 20 years which have reduced housing access chances of rural households rather than the limited influx of second home purchasers. The move towards 'shared ownership' association schemes in the early 1990s and the transfer of this housing to the market (through full-equity staircasing) might be seen as another example of how governments have mis-managed non-market housing in rural Britain and created the current housing crisis. In the light of these policy changes, few of the suggestions made in the prior studies seem relevant in today's political climate. However, in 1983, Shucksmith did note that the cost of a comprehensive investment package in public and voluntary sector house-building might not prove to be the panacea that everybody hoped for whilst the actual cost would be astronomical. At the same time, such a blanket measure could cause the ghettoisation of tenure classes and social polarisation in some rural communities (Shucksmith, 1983, p.191). It seems however, that the Conservative government did not fall into this particular trap; firstly, by not developing a comprehensive financial package; secondly, by allowing the development of rural housing to continue in an *ad hoc* way, and thirdly, by belatedly promoting policies for tenure diversification. Other suggested policy packages, however, did rely heavily on the promotion of council building as a means of satisfying 'local needs' in the face of outside demands on the existing housing stock (Dart, 1977, p.76). These policy packages also pointed to the potential benefit that might be derived from local authorities purchasing, and municipalising, existing local property that came onto the local market (that is, 'competing' on behalf of those who were unable to compete; this issue is discussed below). Despite recent changes in housing policy affecting local authorities and housing associations, it is still clear that any policy measure designed to control the development of second homes (or any purchases by outsiders) must be augmented with some element of public provision in rural areas (Shucksmith, 1985, p.73).

'Municipalising' Empty Properties

The acquisition of empty properties (which can be let to local people), like new-build programmes, has largely fallen prey to Conservative economic dogma. Dower (1977) notes that by 1974, the Labour Government was moving towards encouraging the municipal purchase of vacant properties and development land (the purchase of development land by local authorities was often achieved with the use of compulsory purchase orders before the mid-1980s and the softer policy approach adopted by the Conservatives in the countryside); he added that this strategy could provide a key element in securing additional housing opportunities for rural households (Dower, 1977, p.162). This sentiment was echoed by Jenkin (1985) who argued that when suitable properties came onto the market, the local authority should purchase these properties and let them to families on their own housing waiting lists. The policy could be applied in those areas particularly affected by the growth in the number of second homes in the effective housing stock and could extend to acquiring properties from pensioners in order to alleviate the problem or rural underoccupancy (Jenkin, 1985, p.78). Pyne, in his study of Caernarvonshire in North Wales, conceded that such a strategy would be heavily resource-dependent

(Pyne, 1973, p.47), involving both the acquisition and renovation of existing housing, a move which would not make any net contribution to the overall rural housing stock (arguments later repeated by the Housing Corporation in England and Tai Cymru/Housing for Wales). As early as 1983, Shucksmith argued that the acquisition of existing housing on the part of local authorities (like direct new-build) was no longer a realistic proposition (Shucksmith, 1983, p.185). This direct-handed approach was now the task assigned to housing associations (as the Conservative Government's 'third arm' in its housing strategy) who were devoting a growing proportion of their development budgets to acquisition and rehabilitation throughout the 1980s.

By 1987/88, housing associations in Wales devoted nearly 65 per cent of their aggregate development programme to rehabilitation, but by 1992/93, this figure had fallen to just 4.6 per cent (Welsh Office, 1993, p.37). This substantial decline was largely due to the way in which grant allocations were calculated after the Housing Act 1988. Before this legislation, associations were able to develop new units or rehabilitate acquisitions and subsequently apply for HAG to cover the full costs (within a certain budget limit). After 1988 however, the costs had to be calculated before schemes went on-site. Also, if budget over-runs occurred, then the association would have to meet these added costs though additional private loan sanctions. In effect, because the cost of new-build schemes was easier to predict, the number of 'rehab' projects decreased in the light of the greater financial risks. In addition, it was generally cheaper to build from scratch (and associations were now forced to be more cost-conscious) and Housing for Wales preferred schemes which were seen to be making net additions to the housing stock. In the early 1990s, the future for this type of acquisition and renovation activity looked bleak, however, there is some evidence that this situation is slowly changing. In 1992, the Government established an 'acquisition initiative' in Wales with a budget of £38 million which allowed some local authorities and housing associations to acquire properties from private ownership, repossessions or newly-built properties which could not be sold in the private market. In 1996, the Government established a rural housing 'challenge fund' of £7.5 million in Wales which was intended to be used to promote 'innovative' methods of affordable housing provision which could potentially include acquisition by associations (Welsh Office, 1996a).

Managing Private Properties

Another measure to promote the supply of affordable rural housing has been the municipal *management* of properties remaining in private ownership. The first pilot schemes of this type were attempted in North Wiltshire District in 1976 with private properties being managed (and improved) by the local authority and rented out to local people with particular guarantees and advantages for the owner (relating to the improvement works and the guarantee that the owner can re-gain vacant possession after a set tenancy period). Jenkin (1985) argued that these so-called 'North Wiltshire' schemes could be used as part of a package of measures to combat rural housing shortages (Jenkin, 1985, p.77). A clear disadvantage with this strategy, as Shucksmith pointed out, was the fact that the councils had an obligation to re-house the sub-tenant on the expiration of the sub-tenancy, placing further pressure on the financial resources of the authority (authorities would certainly avoid the situation of having too many sub-tenants and therefore too much potential pressure; Shucksmith, 1983, p.187). More recently, this scheme has been up-dated and 80 local authorities across England and Wales are running 'empty homes' campaigns in partnership with local housing associations. In West Dorset, for example, 7 per cent of properties are vacant (many of these 'empty' properties are second homes) and a 'filling the empties' campaign is trying to ensure that twenty owners each year give up their empty dwellings to local people on waiting lists (*The Guardian*, 26 July 1995, p.29). By mid 1995, the authority had written to 2,353 absent owners and had received 317 replies; the letting of five properties had been agreed whilst negotiation was continuing with a further 24 owners. In 1991, the 'Housing Associations as Managing Agents' (HAMA) scheme was launched and is helping to persuade many owners of the benefits of allowing associations to manage their properties. In return for a small measure of altruism, owners receive grants for renovation work and are guaranteed the vacant possession of the property at the end of the lease. The Empty Homes Agency estimates that 737,000 homes stand empty in the private sector whilst 120,000 families are accepted as homeless by local authorities each year. However, before these 'empty homes' schemes can provide a realistic solution, there is a need to educate owners on the benefits of leasing (many are unaware that assured shorthold tenancies introduced in 1989 allow fixed-term letting) and also to convince rural local authorities that such schemes are applicable outside of towns and cities. The real irony in many rural areas is that a significant number of these 'empty' properties are in fact second homes and it seems unlikely that part of the cause (of some rural housing problems) could become part of the cure. As a general rule, second home owners wish to maintain access to their properties throughout the year and even the 'winter letting' of some second homes is unlikely to have any significant impact on rural housing problems. Whilst seasonal letting to students might go some way towards alleviating the annual accommodation crisis facing students in some areas, the creation of a 'seasonal homelessness' problem in other areas could increase the burden on local housing authorities.

Community Self-Build

In response to particular housing problems in some rural areas, some agencies have advocated the benefits of community self-build schemes. In relation to second homes and the need to expand housing access opportunities for local people, Shucksmith claims that 'community self-build' offers another way forward (Shucksmith, 1983, p.86). The National Federation of Housing Associations published guidelines for people wishing to become involved in low-cost self-build schemes in the late 1980s (NFHA, 1988a, 1988b). It was clear in these guidelines, however, that this particular option was more suited to towns or key rural settlements where larger groups are people are likely to want to become involved (and can share the various legal and material costs). Because of economies of scale, it is unlikely that a scheme involving just one or two households in a small rural village would be able to get off the ground. Clearly, there are logistical and financial difficulties associated with self-build schemes and Shucksmith argues that the 'difficulties of obtaining finance will prevent self-build schemes making any major contribution in areas of high second home ownership' (Shucksmith, 1983, p.186). As Jenkin points out, self-build is never likely to be a viable substitute for local authority or housing association provision (Jenkin, 1985, p.65).

There are a range of innovative housing policy measures which may be useful in alleviating the housing problems in rural areas which may, in some instances, be partially attributable to the growth in second home ownership. In the 1980s, these policies often focused on the promotion of low-cost home ownership. More recently, following the publication of planning Circulars 7/91 and 31/91 in England and Wales (and now Planning Guidance Wales - Planning Policy along with Technical Advice Note 2 'Land for Housing in Wales'), attention has re-focused on the interface between housing and planning policy and the ways in which these two may be synthesised to generate a contribution of affordable rural housing. However, housing policy has not only been used to 'soften' the impact of second home ownership; the prior studies highlight a number of housing-based measures which could restrict or discourage the expansion of second home ownership. Some of these may still be applicable today.

Improvement Grant Restrictions

The first suite of potential measures are designed to increase the overall cost of second home ownership and include restrictions on improvement grant availability (not relevant after 1974), the levying of differential rates (or council tax today), tax penalties for second home owners, a general levy on the purchase price of second homes and restrictions on mortgage advances for second home purchases. Many of the pre-1974 studies highlighted the possibility of restricting access to local authority improvement grants as a means of increasing the cost of second home ownership and therefore reducing overall demand. The rules governing the use of these grants were amended in the Housing Act 1974 (Part VII, s60) and second home owners effectively became ineligible. Shucksmith (1983) has argued that this legislative change (along side the removal of tax relief for mortgages on second homes) was partially responsible for the lull in second home demand between 1973 and 1977 (Shucksmith, 1983, p.176). However, it is interesting to consider whether the removal of these grants contributed to a refocusing of demand on mainstream housing after 1977 as buyers were discouraged from acquiring derelict properties. Similarly, a number of studies in the early 1970s (Pyne, 1973; Jacobs, 1972) showed that the cost of these grants to the local authorities was rapidly recouped in increased rate income. At the same time, improvement work was often a much-needed boost to the local building industry. It appears therefore, that in some areas at least, this blanket measure may have worsened direct competition for effective housing and proved to be a false-economy.

The Tax System

Whilst amendments were made to the way in which improvement grants were managed by the local authorities (for example, through the provisions of the Local Government and Housing Act 1989), the Government turned its attention to the tax system. Pyne (1973) had argued that a key Government option for curbing second home ownership was the imposition of tax penalties (Pyne, 1973, p.40) and in the following year, the Finance Act removed tax relief for mortgages on second homes. The Dart (1977) study in Scotland argued that Capital Gains Tax already discourages those purchasers who might consider a second home as an investment whilst the impact of Capital Transfer Tax and Wealth Tax introduced in the earlier legislation was still unknown (although these tax dis-incentives may have contributed to the decline in demand between 1973 and 1977). In France, Clout (1969) has noted that higher taxes on second homes reflect a recognition of an increased burden on local services (particularly infrastructure). However, in Britain, little public capital expenditure is required to bring a dwelling into second home use and it may be argued that the second home generates an extra use of the relatively underused facilities that are already supplied to permanent residents (Jacobs, 1972, p.48). Whilst the removal of mortgage tax relief may be justified on equity grounds along with the fact that greater levels of consumption should not be rewarded, it is difficult to see how the taxing of ambiguous economic and social impacts might be achieved. The tax system cannot be used to maximise the benefits and minimise the costs of second homes in different areas and therefore this general policy approach is very much a blunt tool.

Differential Rates or Council Tax

The same is perhaps true of the rating (or council tax) system. A number of the prior second home studies have highlighted the possibility of levying higher rates charges from second home owners, again increasing the overall cost of second home ownership. This would be a controversial move particularly as second home owners already pay full rates but only make partial use of local services (Dart, 1977, p.62), a point emphasised by Coppock:

> County studies indicate that these owners of second homes bring financial benefits, contributing more in rates than they consume in services and making a considerable local purchases, in turn creating new employment through the multiplier effect (Coppock, 1977, p.147).

Despite these arguments, it is clear that an additional charge could be levied on second homes, the rationale being that 'increasing cost of the second home would lead to demand for them falling' (Jenkin, 1985, p.58).

105

However, at present, local authorities cannot levy council tax in this way and such a move would require new legislation (to take account of the *use* of those dwellings in the current banding schemes). Such a policy would be discriminatory and would run into considerable opposition for two reasons. Firstly, the policy could not be applied selectively between areas or particular communities; this would mean that all second home owners would be disadvantaged and not only those contributing to local social and economic problems by competing for mainstream housing. In effect, the policy could not achieve the objective of maximising economic benefits whilst minimising social costs. Secondly, great emphasis in market democracies is placed on the freedom of movement and of capital, and in this context, any discriminatory policy would be viewed as an impingement on personal liberty.

Standard Purchase Levies and Restrictions on Mortgage Finance

Two additional ways of increasing the costs of second home ownership would be to apply a standard levy to second home purchase or restrict the availability of mortgage advances. Pyne (1973) argued that a standard levy might be achieved by re-introducing stamp duty (or an equivalent) on selective property purchases including second homes; the additional revenue raised by local authorities could then be used to subsidise new affordable housing projects (Pyne, 1973, p.40). In Ireland, a similar scheme is used whereby foreign purchasers incur additional purchase costs (Downing and Dower, 1973, p.35). More generally, the Government might seek to influence private lenders and decrease the availability of mortgage advances for second homes (Pyne, 1973; Downing and Dower, 1973). Downing and Dower argued that if these measures were applied in moderation, they need not create major political waves. However, it was not clear in the prior studies how the Government might go about influencing lenders and with the present uncertainty over the economic climate, it is unlikely that this strategy would be popular with high street banks and building societies who would certainly reject any moves adversely affecting the demand for loan capital.

Property Licensing and the Creation of Housing Sub-Markets

Altering the financial framework in which the demand for second homes operates offers an opportunity for indirectly affecting the housing market. There are, of course, more direct policy options available including the general introduction of property purchase controls (that is, 'licensing' purchases) or the creation of exclusive property sub-markets. The first of these options is already in operation in the Channel Islands and in the Netherlands (Pyne, 1973, p.46). Since 1945, Jersey's Housing Committee has been charged to consider all applications of property purchase and land use change; similar measures have been applied in Guernsey since the Housing Control Law of 1969 (Williams, 1974, p.60). Only potential purchases meeting specific residency and employment criteria are granted the license to acquire property. However, a number of arguments have been voiced against the extension of such a scheme on the British mainland including the negative economic impact of licensing in some importing regions and the more general effect on the balance of payments as second home buyers seek properties abroad (this is a wider argument against all prohibitive measures). Practically, opponents argue that such a scheme would be extremely difficult to administer over a wide area although this counter-argument is perhaps less damning than the likely impact of such a scheme on personal property rights and the individual's freedom of residence.

The essential point in 'licensing' schemes is that control could be exerted over the change of use of *existing* dwellings (licensing could be implicitly used to bar second home seekers from the market because they fall short of residency or employment requirements). This cannot be achieved with the use of current planning legislation where it is only new properties which may be subject to development control or occupancy clauses. As with licensing, the creation of property sub-markets (from which certain consumption groups are effectively excluded) has been proposed as a means to protect 'local' interests. In Wales, the Welsh Language Society continues to argue that 'the private purchasing market must... be controlled so that local people can purchase property at a reasonable price' (Cymdeithas Yr Iaith Gymraeg, 1989, p.1). In the face of rising property prices (this particular document was published at the height of the property boom) in rural areas, the Society argued that the only answer was to control prices and access to existing housing stock with local people given priority when bidding for new and old housing. For cultural, linguistic and nationalistic reasons, the Society seeks to prioritise local interests despite arguments that any and all local housing initiatives 'unfairly advantage those on low incomes who, by good fortune, live in rural areas' (Hutton, 1991, p.311). The Society proposed that the District Valuers Department in each local authority should set 'fair prices' in the area and that a 'points system' should be extended to the purchase of private property with the principal component being local residence. In practice, someone wishing to sell a property would have to contact the local authority; if there was a need for more rented accommodation in the area then the authority (depending on resources) would purchase the property and let it to local people. If this was not the case then the property could be sold but not with a 'market' value. Instead, a 'banding system' would be derived whereby local first time buyers have first refusal on the property, people from adjacent areas have second refusal and only if there are no buyers in this local needs 'cascade' could the property be advertised further afield (with price restrictions). The policy's designers claimed this level of control and the creation of such a rural housing sub-market offered the 'only long term answer to this worsening [housing] problem' (Cymdeithas Yr Iaith Gymraeg, 1989, p.7). Clearly, the policy takes a very narrow view of both the rural housing 'problem' and a very basic view of the way in which rural housing markets operate. This type of scheme is politically unattainable; from a development point of view, this radicalism occupies a polarised position in the field of policy options and echoes Plaid Cymru's sentiments in the 1970s that tourism and second homes are an unmitigated disaster for Wales and for the Welsh (Coppock, 1977).

Increasing calls for a 'property act' in Wales during the late 1980s and at times in the 1990s in some ways represent a knee-jerk reaction to the wider housing crisis facing rural households across Britain; a crisis that was accentuated by the property boom and the rush to buy homes caused by the deadline on the abolition of multiple mortgage tax relief on single properties in August 1988 (Drabble, 1990, p.5). Despite the obvious impracticalities and shortcomings of the radical response, it is becoming increasingly clear that some means must be found of diverting attention away from some *existing* rural housing. It has been shown that second home seekers are increasingly drawn to the effective housing stock as the supply of surplus housing evaporates. By out-bidding locals in some areas, outside buyers are unwittingly contributing to the growing problem of rural homelessness and putting additional pressure on local authorities and housing associations to increase building programmes where this may not otherwise have been necessary. Second homes are not a problem *per se* in the countryside but simply represent part of the evolution of the housing market in response to economic decline and social change. However, they may become a problem where the competition for existing mainstream housing leads to the displacement of permanent residents. This is the crux of the second home problem and requires a policy response. Many observers believe that the land use planning system may hold the key to a solution.

Planning and Development Control

The potential for using the land use planning system to control the growth of second home ownership and minimise negative impacts is discussed in length in Chapter Eight. The purpose of the discussion here is simply to introduce the more salient issues. The final point in the last section demonstrated that there is a clear separation between controlling new development (and its occupancy and use) and influencing the occupancy and use of existing housing. Arguably, whilst controlling new development is the simpler course of action, influencing the use of existing housing would bring far greater benefits for many rural communities affected by second homes.

New Development and Occupancy Controls

In terms of new development, occupancy may be controlled by specific local needs policies written into development plans, planning obligations or planning agreements. Local needs policies may indicate that in certain communities, new housing development will be limited to providing for proven local needs. The 'rural exceptions' strategy in Wales is outlined in TAN(W) 2 'Land for Housing in Wales' and illustrates how off-plan land may be developed (where a landowner chooses to release that land) to provide affordable housing for local people. Similarly, the planning authority may specify sites where private development can occur but only if it includes an element of local needs housing. Some local authorities believe that they have a free-reign to implement rather explicit local needs policies. Ceredigion District Council, for example, attempted to implement the following policy through its local plan:

The occupation of new dwellings, including the conversion of existing buildings, shall be limited to persons and their dependants who originated from or have lived for a period of five years from the date of occupancy either in Ceredigion or in an area no more than twenty-five miles from its boundary.

The Welsh Office questioned the legitimacy (although did not intervene) of some of these policies, arguing that the planning system should be only concerned with land use and not the merits of the user. Neverthleless, Ceredigion did withdraw this policy following adverse publicity. However, the much-cited 'Mitchell' case brought before the Court of Appeal in 1993, seemed to confirm that the need for social housing was a legitimate planning consideration and clearly does relate to the type of user to occupy a new dwelling. Further official support was voiced for local housing policies in the Department of the Environment Circular 13/96. This Circular confirms that the need for affordable housing is a material planning consideration and gives further weight to local authorities negotiating an element of social housing in sale housing schemes.

These policies can, therefore, be used to ensure that some of the needs of rural households are met by new development. Additional planning obligations and agreements complement these policies, ensuring that housing is retained for local needs in perpetuity (and is not sold-on to outside buyers). Obligations have proved fairly ineffective at controlling occupancy as they are often overturned by appeal. Agreements, on the other hand, particularly those under Section 106 of the Town and Country Planning Act 1990 have proved more successful in controlling occupancy and are widely used on rural exception schemes. Their predecessor, the Section 52 Agreements (from the Town and Country Planning Act 1970), have been studied in detail by Shucksmith (1980, 1990a) who argued that where the amount of new development is restricted to meet local needs, reductions in housing supply result in further increases in house prices, accentuating problems in the market for existing properties (Shucksmith, 1981, p.142). The general conclusion tends to be that control over new development is a semi-solution which, at the end of the day, amounts to no solution at all. Where second home purchasers play a significant role in the housing market, competition for existing housing increases whilst the access chances of local low income, low wealth groups decrease. It is in this situation that opposition to second homes flourishes.

Existing Housing and Change of Use Controls

In this context, there are two courses of action available. Firstly, directly controlling the property market either by licensing purchase transactions or creating sub-markets in which prices are regulated. As this option is unlikely to attract a great deal of popular or political support, the alternative is to bring *change of use* of existing dwellings (from first to second home use) under planning control. This is not a new idea and has been advocated for over the past twenty-five years (Bielckus, *et al*, 1972; Downing and Dower, 1973; Dower, 1977; Pyne, 1973; Jacobs, 1972; among others). The trend towards second homes being drawn from the mainstream housing stock through change of use in the 1970s led many observers to call for new legislation making such a use change subject to planning permission (this may have been possible under the provisions of Section 22 of the Town and Country Planning Act 1971). In 1981, Dafydd Wigley MP introduced a private member's Bill to this effect. In response, the Conservative Government rejected the Bill, arguing that the use to which an owner puts a property (which is still residential) is not a material planning consideration; in effect, they argued that disqualifying people from owning second homes in this way would amount to an infringement on personal liberty. At that time, it did seem that the policy would 'undoubtedly run contrary to the spirit of existing planning law' (Jenkin, 1985, p.57). Arguably, this 'spirit' has now altered and the planning system operates in a less mechanistic way and is more willing to pay lip-service to the needs of the local community (again, this is underlined by Circular 13/91). Whilst Dafydd Wigley's private Bill failed in 1981, the Labour Party expressed support for change of use legislation, committing themselves to introducing their own Bill when they were asked to form a Government. Currently, change of use legislation would require an amendment to the Use Classes Order 1987 and in 1990, the Labour Party again expressed support for such a move (Soley, 1990, pp.38-39). Although the Conservative Government in the 1990s maintained that this policy response is 'spiteful and impractical', the Labour Party's success at winning the 1997 General Election has not transposed the pre-election second homes thinking into policy responses. Nevertheless, the second homes subject is once again entering into the political arena. The change of use option may once again offer an opportunity that could be implemented through amendment by an elected assembly or executive at the local level, to deal with the more detrimental impacts of second homes apparent in some areas.

The range of policy options examined in this Chapter are summarised in Figure 3 overleaf.

Figure 3: Policy options for the UK

Policy measure	Legislative change	Implementation	Examples	Comments
Affecting housing supply and the economy				
Economic growth:				
1 Local strategy	Yes: changes in EU Structure fund status could benefit some regions	Local and National	-	The long-term solution to control the underlying causes of rural decline; pro-active second home development could be an element of such a strategy
2 National strategy				Tackles the key issues of rural housing supply
Assisting locals through non-market housing provision:				
Direct provision	Yes: improved funding	Local	-	Inadequate provision (includes shared ownership)
Social sector acquisition/rehab	Yes: improved funding	Local	-	Likely to make a larger contribution in towns
Managing private housing	Yes: improved funding	Local	-	Empty rural properties are often unfit or second homes
Community self-build/other voluntarism	No	Local	-	Inadequate provision and more suited to towns; is a response to government inaction
Affecting housing demand and consumption				
Raising the costs of second homes:				
General taxation	Yes	National	France	Does not differentiate types of second homes and may penalise locals (as in Sweden)
Differential Council Tax	Yes	National	-	As above
Standard purchase levies	Yes	National	Republic of Ireland	As above
Controlling the property market:				
Property licensing	Yes	National	Scandinavia/Channel Is	Geared to the issue of foreign purchasers
Housing sub-markets and property rights	Yes	National	Finland (Åland)	As above; can only operate in areas of devolved government where local interests take precedence over national concerns
Planning:				
Occupancy controls	No	Local	United Kingdom	Control the occupancy of some new housing, but legally fraught
Change of use controls	Yes	Local	Denmark	Controversial and difficult to implement

8　The British Planning Policy Process

Planning has long been chastised for its failure to focus on broader social concerns rather than narrow land-use issues. Even today, the issue of affordable housing provision in rural areas remains a contested subject for the planning process to accommodate in the UK (Gallent, 1997), and one within which legal contentions and political debate continue.

Within certain parts of the UK, particularly Wales, the statutory planning system's narrow focus has been a source of bitterness amongst rural local planning authorities and community and linguistic campaigners. On occasions when local decision-makers have attempted to take decisions that meet community needs - but which are based on social and linguistic considerations, rather than strictly on land use criteria - Central Government has accused those local planning authorities of malpractice in their use of the planning system (for example, see House of Commons Welsh Affairs Committee, 1993; Tewdwr-Jones, 1995; Cloke, 1996). This frustration in rural communities in Wales has centred primarily on rural housing provision in those areas that possess a high proportion of Welsh speakers, and the planning system's failure to address rapid socio-cultural change, high levels of in-migration by non-Welsh speakers, the on-going decline in agriculture, perceived threats facing the Welsh language, and the lack of means through which affordable housing can be procured for local people.

The inability of the planning system to respond to broader social problems in rural areas has resulted in high levels of political and community frustration, and the emergence of opposition to the way in which the statutory planning process should be operated. In Wales and the South West of England, for example, certain local planning authorities have attempted to base their decisions upon the personal circumstances of the applicants for planning permission, rather than considerations relating solely to the use and development of land. The lack of broader mechanisms within the planning system to ameliorate wider rural problems (including concerns for the language, the lack of affordable housing and the issues surrounding in-migration and second home purchasing) have contributed to a belief in those communities that planning simply exists to benefit non-local property classes.

We commence our discussion of the planning policy processes with an examination of the mechanisms that could be utilised to both contribute to the amelioration of the negative image associated with the second home issue and indirectly to rebuild the shattered image of the planning system in some rural areas of the UK. The discussion is centred on three issues: the use of occupancy controls; amendments to the Use Classes Order; and recent political developments. This Chapter focuses attention on how planning control may presently be used to regulate the occupancy (and therefore use) of *new* housing and how planning control may, in the future, be used to restrict the number of *existing* dwellings becoming second homes. It is not the purpose of this Chapter to consider, in detail, the way in which purpose-built second homes (either as individual units or larger grouped developments) should be planned, nor is it intended to provide an in-depth examination of what is and is not possible within a particular local area. However, this issue is touched upon in the final section which examines the more holistic policy responses within a UK context.

Background to the Use of Occupancy Controls

In the 1970s, planning agreements were often used by local authorities as a means of securing 'a contribution from developers towards the infrastructure costs' of new housing schemes (Barlow, *et al*, 1994, p.3). These agreements were commonly based on the provisions of Section 52 of the Town and Country Planning Act 1971. By the mid 1970s, some local authorities were using such agreements with developers to procure new council housing units as well as other social infrastructure. Critically, Loughlin (1984) revealed that authorities were also using these agreements (and other conditions) to limit the occupancy of new dwellings in some rural areas to local residents. This phenomenon had already been studied in detail by Shucksmith in the Lake District (1981). Where planning agreements are used (often as part of local needs policies written into development plans), the control of occupancy is closely linked to the actual development process. This discussion however, focuses solely on occupancy control rather than the procurement of new affordable housing, the subject of numerous research reports in recent years (Joseph Rowntree Foundation, 1994; Barlow and Chambers, 1992; Barlow, *et al*, 1994).

The Use of Occupancy Controls

By 1985, more than 20 structure plans contained references to specific 'local needs policies'; that is, an explicit commitment to securing planning gains, through negotiation with developers, which emphasised the overriding importance of local housing needs (Bishop and Hooper, 1991, p.13). These policies invariably contained references to the control of occupancy. In 1981, Shucksmith argued that the housing problems faced by some rural communities could be alleviated in three ways; by discouraging or preventing houses from being used as retirement or holiday homes, by building more council houses and increasing rental stock, or through planning control with new dwellings going to local people. The Lake District Special Planning Board's (LDSPB) use of Section 52 agreements was examined in detail. Shucksmith found that the control of occupancy increased the competition for existing housing stock and caused further elevation of house prices. The experience of the LDSPB offered clear evidence that it is impossible to restrict the occupancy of new dwellings whilst responding to local needs (Shucksmith, 1981, p.141). This view was rejected by the Secretary of State for the Environment, Patrick Jenkin, when he deleted the use of planning agreements to control occupancy from the Cumbria and Lake District Joint Structure Plan in 1984 (Jenkin, 1985, p.56). If prices rise in the existing housing stock (because of the planning restrictions placed on the occupancy of new dwellings), then an increasing proportion of this stock will be unavailable to low income, low wealth rural households. The access chances of this group, therefore, will be increasingly dependent on the amount of new affordable housing being developed.

There is evidence to suggest that affordable housing provision is insufficient to meet growing need. The shortfall in housing provision nationally is usually estimated to be in the region of 100,000 units per annum in England (See Wilcox, 1990; Whitehead and Kleinman, 1991; Audit Commission, 1992) and 22,000 per annum in Wales and Scotland (CIH, 1992). In rural Wales, Bramley concluded in 1991, that 'the deficiency [in provision] is dramatic where the need for social housing is 10 times the current building programme' (Bramley, 1991). Even though the joint efforts of local authorities and associations led to the completion of 31,558 dwellings in England in 1994 (HM Government, 1996), research by the National Federation of Housing Associations (1994) suggested that the shortfall was still in the region of 124,000 units. In Wales, there were just 2,797 public and voluntary sector completions in the same year.

As a consequence, it is possible to suggest that there is insufficient new housing being developed to meet rural needs. Occupancy controls (achieved through planning obligations or the use of Section 106 agreements from the Town and Country Planning Act 1990, the successor of Section 52 agreements) may mean that local people can monopolise new housing in some areas, but there may be serious side-effects. Firstly, new affordable housing becomes increasingly associated with low-income groups and a social stigma may become attached to these developments, particularly where schemes are grouped together to form social housing 'ghettos'. Secondly, these controls may inhibit the ability of associations to attract private finance on exception schemes, further reducing access opportunities for rural households (see Rural Development Commission, 1995). Thirdly, and perhaps more significantly in the current debate, occupancy clauses offer no route into the existing housing stock; indeed the exclusion of wealthier groups from new housing means that market competition concentrates on a smaller stock of existing dwellings and therefore price rises may be more acute than they might have been if all groups could compete for all housing, existing and new. The purpose here is not to suggest that occupancy controls do not have a legitimate role to play in alleviating rural housing problems. Rather, the point is that the control of *new* property *alone* may accentuate more general problems in the wider housing market.

These problems may become more pronounced in the years ahead given the increasing support for occupancy controls as an element of affordable housing policies. Before 1989, it was widely accepted that the planning system could only legitimately deal with land-use issues and that in the case of housing this excluded considerations of tenure, price and occupancy. Clearly, this system should not be used to promote particular tenure classes; control over occupancy was frowned upon. It is difficult to judge how far the present system has moved away from this position over the last ten years although the debate surrounding the 'need' for social housing as a 'material planning consideration' has not yet been fully resolved. The Department of the Environment Circular 13/96 and the Department of the Environment Transport and the Region's Circular 6/98 both confirm the overall legitimacy of affordable housing policies and it is clear, that in some circumstances, occupancy control is vital. In the case of rural exceptions policy, for instance, it is unlikely that the altruistic stance of landowners would be maintained if it was thought that dwellings built on land released at below development value would be quickly sold on at market value.

Occupancy controls are necessary in certain instances, but their social benefits may be outweighed by socio-economic costs if they are used too freely. In Ceredigion, for example, moves to control the occupancy of all new dwellings along with conversions of existing dwellings is likely to have a serious impact on house prices in those communities where the housing market is affected by a significant number of outside buyers. However, the councils employing such policies should not be ridiculed for their attempts to prioritise local interests (afterall, they are the elected representatives of these interests). The means to achieve more balanced control across the entire property market are simply not available. If they were, this might result in the creation of more balanced communities (with different consumption groups occupying a greater range of housing types and tenures), the suppression of excessive demand and the relief of growing social tension.

Political Developments and the Prospect of Change of Use Controls

Part I of this book argued that the creation of housing sub-markets is not a viable option in England and Wales where, for the present time at least, the "national" perspective prevails and the maintenance of property rights across the entire market is seen as both practically and ideologically essential. There is a way, however, to maintain the integrity of the market and market allocation, and at the same time, control the occupancy (or rather, the *use*) of existing dwellings. This could be achieved, as in Åland, by drawing a planning distinction between property used for a primary residential function and secondary residential property. This distinction would be retrospective and all existing property would be classified in this way (in Åland, the *use class* is defined when new dwellings are constructed). In effect, *change of use* would subsequently be subject to planning permission (and controlled by local authorities); that is, planning permission would be required for a dwelling used as a first home (for the last 5 or so years) to become a second home and vice versa. The demand for second homes would be suppressed and the value of properties might decline. However, it is true that all planning control may have an inflationary or deflationary effect on property prices and therefore, in this respect, opposition from existing owners would probably be limited. Despite the criticisms of occupancy controls in 1981, Shucksmith argued that:

> Planning controls [focusing on use change] are potentially the most sensitive means of reducing the number of second homes, in that the policy would be operated by local authorities who could apply such restrictions in accordance with local circumstances (Shucksmith, 1983, p.189).

For instance, permissions might be granted where properties were surplus to general housing needs (and where the benefits of second homes outweighed the costs) but refused where change of use from first to second homes was causing the classic symptoms, of house price inflation and social disruption.

The concept of bringing this type of use change under planning control is not new and has been canvassed by a number of academics, planning practitioners and politicians for nearly 30 years. In the early 1970s, it was recognised that the distribution of second homes in England and Wales was the product of dwelling use change; this had been true for a number of years, but increasingly, the balance between surplus stock and mainstream stock use change was shifting in favour of mainstream stock with the associated economic and social problems. At this time, a number of observers commented that this change of use should be brought under planning control (Bielckus, *et al*, 1972; Downing and Dower, 1973; Dower, 1977; Pyne, 1973; Dart, 1977) by amending Section 22 of the Town and Country Planning Act 1971 (the modern equivalent would involve defining a new use class in the 1987 Use Classes Order). By the 1980s, the idea had reached the political agenda with the introduction of Dafydd Wigley's Private Member's Bill in 1981. At that time, a number of criticisms were levelled at the possible creation of an additional use class. The Conservative Government's argument hinged on the potential 'infringement to personal liberty' brought by such legislation. It could be argued, however, that given the way in which that Government reduced the supply of rented accommodation (and infringed on consumer choice) and its apparent approval of occupancy clauses, any such criticism of amending the Use Classes Order today might appear misplaced. In the early 1980s, it was clear however, that the type of material use change envisaged focused on the merits of the user and ran contrary to the spirit of planning law. Again, the context today is much changed and there is perhaps little reason to believe that a system which allows the control of dwelling occupancy could not stretch to differentiating between *types* of residential use. Another problem highlighted at the time of the Wigley Bill was the fact that a particular housing right might be overridden by new legislation (that is, the right to use residential property for residence, either temporary or permanent) and therefore compensation, running into many millions of pounds, might be payable to property owners unable to exercise their right through change of use.

Despite Conservative objections, instrumental in the failure of the Bill during its first reading, the concept of change of use as a means of regulating second homes in the existing rural stock received a warm reception in Labour Party ranks. At the time, the Labour Party expressed support for the policy and said publicly that they would introduce such a measure, as part of a new planning package, when they were returned to Government. The political pledge made in 1981 regarding amending the planning system was clarified in

1990. At this time, the Labour Party argued that reductions in public subsidy for social housing revealed that the Government was unable and unwilling to tackle rural housing problems. In contrast, Labour claimed to have a policy package which would create sufficient funding for the voluntary sector and reform housing finance in such a way as to bring home ownership within the grasp of a great many more households. One element of their rural strategy concerned second homes, although the Party recognised that this phenomenon was in many respects only symptomatic of underlying socio-economic forces. Soley (1990) outlined the Labour position on second homes in the planning context:

> Our proposal is that local authorities involved would be able to restrict the growth of second homes in the affected areas by ensuring that an existing family home would need planning consent for *change of use* before it could be sold as a second home in much the same way as change of use is required to change a home into an office, for example. The procedure would be subject to appeal. This is the practice in a number of European countries. There is no question of a ban on second homes (Soley, 1990, p.39).

In the Labour model, the policy would not be applied retrospectively (that is, it would involve all dwellings, but recent use changes would not be subject to re-appraisal) and would not affect those people who require two homes for their work, those who are planning to retire to an area and those who live in tied accommodation. Continuing Conservative criticism that such a policy move would be 'spiteful' was rebuked by Labour's policy designers who, again, claimed that this attitude 'smacked of hypocrisy' given the Conservative Party's track-record in reducing housing opportunities for rural people. To some extent, the Labour Party did try to win support for the policy by tapping peoples' nostalgic image of the countryside. In 1990, Soley argued that 'our villages are a very important part of our heritage and should not be allowed to die on their feet' (Soley, 1990, p.39). Since their election to Government in May 1997, the Labour Party has shied away from the second homes issue and the policy itself was not apparent within the pre-election political manifesto. While it remains relatively 'early' in the New Labour parliament, there remains scepticism as to whether the Government will eventually turn its attention to the second homes phenomenon, given concern amongst countryside and rural interests at the perceived urban bias of the Government's policies.

In the late 1990s, the second homes debate once again returned to the political agenda at the national level in the UK. In April 1998, Elfyn Llwyd, the Plaid Cymru Member of Parliament (MP) for Meirionnydd Nant Conwy in North Wales introduced a Private Member's Bill in the House of Commons in

120

Westminster aimed at amending the planning legislation in order to protect rural villages in Welsh-speaking Wales. His Bill sought to introduce three measures: to impose a time limit of five years on planning permissions awarded by local planning authorities where the develop had not yet started any substantial development; to introduce a third party right of appeal for residents, objectors and others interested in applications for planning permission; and finally to introduce amendments to the Use Classes Order 1987 to classify all housing as primary or secondary residences with a requirement for planning permission to change a use from one use to the other. This latter proposal reflects the most likely way academics have suggested the second homes issue could effectively be tackled in the UK. As the MP remarked in introducing his debate into Parliament:

> Such properties [second homes] also bring certain social problems: the usual scenario is that they are purchased by those who can, by definition, afford a second home, which often means that they are earning many more times the average salary of those in the local community. They are able to pay the asking price without quibble, and locals are left in the wake wondering when, if ever, they will be able to enter the property market, and buy a house in the community in which, often, they were born and brought up (Elfyn Llwyd MP, Hansard, 28 April 1998, column 148).

The Bill received political backing at the first reading and was scheduled to receive a second reading in the House of Commons in July 1998. However, this Bill was "lost" when the House was unable to progress all matters during the allotted day for debate. As a Private Member's Bill, it was not surprising that it did not go through to Royal Assent, and the Plaid Cymru Members were aware of the limitations. However, they were enthusiastic for the issue to receive the maximum amount of publicity through media reporting, and vowed to reintroduce the Bill either in a subsequent Parliamentary session at Westminster, or in the National Assembly for Wales.

The Welsh MPs were not the only politicians raising the second homes matter in 1998. Further media reporting was accorded to the Lake District in Cumbria, North West England. The Lake District National Park (LDNP) authority (and its predecessor) has always been proactive in probing the boundaries of the planning legislation in order to increase affordable housing provision through occupancy restrictions, whilst limiting second home development in the rural villages of its administrative area. The LDNP's most recent foray has been to introduce policies within its local plan to permit new housing development only where it would be sold to local people who live or work in Cumbria. Prospective second home purchasers would therefore have to bid for the existing housing stock. The second home phenomenon in the Lake District has escalated in recent years and is now in numbers on a par with parts of North Wales. 15 per cent of all homes in the Park are second homes, and in the villages of Skelwith Bridge and Patterdale, second homes comprise as many as 40 per cent of all housing.

As an additional proposal, South Lakeland District Council announced that they would raise Council Tax levies by 50 per cent, to 100 per cent, on second home properties, thereby raising £1.7million per year in local revenue, but also indirectly acting to discourage the less determined second home seekers from purchasing properties. The Park's plan was put forward - uniquely - by the local planning authority as a Private Member's Bill in Parliament in early summer 1998, and if it had been approved would only have applied to the Lake District.

In both the Lake District and the North Wales Private Member's Bills, the second homes problem lies at the heart of the proposals. This indicates a determination at the rural community level to attempt to introduce measures to ameliorate the problem locally through nationally-legitimised means. Both regions also exhibit distinctive socio-cultural reasons for promoting second home solutions; in the Lake District, the second home purchasing levels reflect the tourist pressures in the area, associated with the picturesque landscape and outsiders' desire to experience something of the "rural lifestyle". In North Wales, the tourist pressure is equally evident but is also compounded by fears over the decline of the Welsh language and the lack of affordable housing for local people. Neither proposal succeeded in Parliament due to insufficient time, but both campaigns attracted significant media attention since they highlighted rural frustration at a time when the present government has been accused of ignoring the needs of the countryside (see Rowe in *The Independent on Sunday*, 10 May 1998).

Finally, there is one further political mechanism that could be employed in the UK to tackle the second homes problem, albeit in Wales and Scotland only at the present time. In May 1999, the National Assembly for Wales and the Scottish Parliament commenced sitting. The Scottish Parliament possesses both statutory and policy-making functions, while the Welsh Assembly does not possess legislative powers - Acts of Parliament will still be debated and passed at Westminster and apply to both England and Wales - but will possess policy-making duties and an ability to amend, if it so desires, secondary legislation. Amendments to the Use Classes Order in the form of primary and secondary residential classifications could therefore be made by either the Scottish Parliament or the National Assembly without requiring Westminster approval. The amendment could be made in reflection of the distinctive problems existing within the two countries with regard to language and socio-cultural pressures, but only if the politicians in the Parliament and Assembly recognised the necessity to act.

However, politics aside, it is the practicalities of the policy which would be crucial if a future Government decided that a policy change was desirable.

Change of Use in Current Legislation

During the evolution of the use change debate, the planning system has been significantly overhauled and it is necessary to consider how such an amendment would fit into the current system. Clearly, some of the changes occurring may have increased the pressure for and viability of a future change to the Use Classes Order. In particular, the increasing recognition by the British courts of affordable housing needs as a material planning consideration has moved the debate on at a positive pace over the last few years, and this judgement has assisted many rural local planning authorities to develop local plan policies for the development of affordable housing where a clear need has been demonstrated and justified by the authority through, for example, local needs surveys. Since the late 1980s, planners and academics have advocated reform of planning law and the introduction of a *social* housing use class (Shucksmith and Watkins in ACRE, 1988) in the Use Classes Order 1987. However, the failure to attract widespread support for this idea hinges on the fact that the Use Classes Order is not designed to be a mechanism of favouring particular types of occupants. Instead, it is 'intended to be an instrument for allowing changes of use which would constitute development were it not for the Order' (Barlow, *et al*, 1994, p.5). The insertion of 'primary' and 'secondary' residence functions would appear to be well within the scope of the Use Classes Order as this move would not necessarily mean control over new development. In the Danish model, however, it means just that; planning permissions are granted to new development depending on the use which is to be made of new dwellings. Therefore, local planning authorities wield control over both new second home development and change of use. In Britain, this type of mechanism would fall at the same hurdle as the social housing use class with criticism that this type of development control is outside the legitimate purpose of the Use Classes Order. In order to overcome this problem, all new development would need the same 'residential classification'; the sub-classification would then be made only *after* the use became apparent (in Denmark, housing used as a primary residence for five years is subject to use change control) and therefore the Use Classes Order could not be abused by local planners.

However, the idea of a *social* housing use class has been rejected on the grounds that 'it would lead to excessive general interest by planners in the personal circumstances of occupiers of housing' (Barlow, *et al*, 1994, p.5). Clearly, this criticism could apply equally to second home use control despite the fact that planners would have no influence over the initial use made of a new dwelling. Where existing dwellings are concerned, 'change of use' control would amount to an invitation to control the occupancy of housing by reference to the circumstances of users; in effect, higher prosperity groups would be excluded from the housing market unless they were intending to move into an area permanently. Another issue is the problem of distinguishing 'permanent' and 'temporary residence'. In Sweden, a large proportion of 'second homes' are occupied for a large part of the year by the wives and children of businessmen who are employed in the cities and live in urban apartments which are 'officially' the family's first home. The distinction, however, is blurred and it is almost as if the rural residence is the first home of half the family whilst the urban apartment is the first home of the working partner. The issue here is how easy and practical it would be to draw clear divisions between primary and secondary residences for planning purposes. In 1990, the Labour Party claimed that the purchase of 'second' homes would not be barred where the user needed the dwelling for employment purposes (Soley, 1990, p.39). Increasingly, more people do work in the countryside. The Conservative Rural White Paper for Wales (Welsh Office, 1996a) claimed that 'geographical location' is becoming 'irrelevant' in the new information age (Welsh Office, 1996a, p.1). This means that many more people, with access to computers and the Internet, can work as easily from home as they can from an office in Manchester, Cardiff or London. In effect, people may have two work locations; one in or close to the city and another far removed, perhaps in a rural backwater, where individuals are able to undertake "telecottaging". But which is the primary residence and which is the second home? If the cottage in the country is no longer used for 'mainly recreational purposes' and is more than an 'occasional residence', can it still be classified as a second home or does the household now have two residences with no distinction between primary and secondary? If the legislators at the national level (and the local planning authorities) are unable to resolve this dilemma, then new legislation will either be generally ineffective or a large number of use change refusals will be overturned on appeal (and prove to be administratively and financially complex).

It is likely that new legislation will not be water-tight and some 'traditional' second homes will slip through the net. However, many would-be second home owners will be deterred by the new legislation and it is likely that use control will have a significant impact on the demand for dwellings in the existing housing stock (and of course, the demand for purpose-built second homes will rise). However, national legislators, wary of giving local planners too much control over occupancy and the housing market, would probably impose tight restrictions on the use of any new legislation. It is likely, for example, that the use of new controls would have to be specified in the local development plan and only used where there was a proven need to ensure that housing did not change use. This 'need' might be based on an assessment of the consumer thresholds of local services, the need for housing for locals or proof that 'tourist accommodation' (development subject to holiday occupancy conditions) in the area was already at an optimum level. These assessments, however, are likely to be very arbitrary with different authorities using different methods for measuring the indicators (this already happens with housing needs assessments). This process itself is likely to be controversial and disputes may frequently be played out in the courts. Many authorities will be keen to demonstrate the social disruption caused by second homes in the local housing market and may even overplay the social costs; other authorities, keen not to inherit the overspill of demand from their neighbours will be forced to play the same game and over time, with more authorities clamping down on use change, rural property prices may suffer a general decline. On the one hand, this might benefit local households who suddenly find themselves in a far stronger competitive position in the rural housing market. On the other hand, the previous revenue from outside investment in the rural property market now stays in the exporting regions; marginally more money may be spent in rural guesthouses and hotels, but most will be invested in leisure pursuits either at home or abroad (affecting the balance of payments). The suppression of rural property prices (and a widening rural/urban differential) may generate new demand for permanent rural dwellings (particularly in the 'information age' when people can readily work from home) causing a total, and permanent, socio-cultural re-configuration of the countryside and further affecting socio-cultural circumstances in particular areas.

The above description of what could happen were use controls imposed on the rural property market is very much speculation. However, the possible adverse effects of new legislation need to be examined urgently given the recognised inadequacy of occupancy controls over new development. Local authorities need to consider the overall effects that over-zealous implementation could have; they should recognise that the way in which they operate may affect both neighbouring authorities and the wider housing market. Any use change legislation would have to deal with establishing a workable distinction between secondary and primary residence. However, any distinction is likely to have its limitations and many households will still be able to acquire what, to all intents and purposes, appear to be second homes. Legislators will have to ensure that local authorities use new controls in a sensitive way and this would inevitably result in a greater monitoring role on the part of elected assemblies and executives and possibly local government ombudsmen.

Given the likely charged political context and complex legal framework within which a use class classification would operate, it would be advisable (and probable) for the any Government amending the use class legislation to design a fast-track planning appeals system in order to mediate in second home cases. Such a system could be used where justified as a way of overturning local authority planning refusals where it is clearly demonstrated that the aspirations of second home owners (and thus allowing second home development) would have no land use impact - in the broadest sense - on the local community. Part I demonstrated that assessments of the costs and benefits of second homes are highly subjective and in this context, deciding whether or not to refuse a planning permission may prove extremely difficult, and costly. Owners are always likely to forward some planning grounds for appeal unless the local authority's case is legitimate and water-tight. A local authority's case in this context would have to operate within the established planning parameters, according to the use class legislation, and be subject to conformity with relevant national and regional planning policies.

Change of use legislation might be effective in alleviating the problems of excessive second home ownership in the existing housing stock in some communities, so long as the legislation was used sensibly and was not allowed to degenerate into a witch-hunt. National legislators would understandably be wary of handing over new powers to local authorities who could be perceived by elected assembly legislators as already operating 'on the fringe' of current legitimate planning controls. The ability to suspend use-change powers (on the part of the elected assemblies and/or executives) might be a prerequisite to the establishment of any new legislation.

Utilising Land Use Planning to Restrict Second Homes

If formal amendment to the Use Classes Order is viewed as inappropriate or else excessively difficult to implement, could housing use be controlled through other planning mechanisms? It is noted, with interest, that planners (urged on by the Welsh Office in the mid to late 1980s) did take an interest in the occupants of new rural dwellings in their implementation of Welsh Office Circular 30/86 regarding housing for senior management (Welsh Office, 1986).

This Circular (which introduced another material consideration in the assessment of planning applications and had to be taken into account by authorities when drawing up their development plans) stressed the Government's desire for planning authorities to make land available for the development of housing suitable for *senior management* and *executives*. The planning justification for this requirement was economic: it was intended to facilitate inward investment by supplying housing of an appropriate standard and scale sought by senior executives. It is somewhat ironic that, at a time when the rural affordable housing question was causing grave concern to local planners, the Government was making a planning commitment to facilitate high-priced housing for particular groups, by encouraging the release of land for housing purposes in or adjoining existing settlements and villages that would not normally be earmarked for development. The argument that planners should not take an excessive general interest in the personal circumstances of occupiers of housing has therefore been undermined in the past by Central Government policy.

Although Circular 30/86 is now cancelled and the senior management housing policy requirement has been deleted from national planning policy in Wales, it does indicate that housing occupation can be a material consideration when it is justified on land use grounds. The same arguments apply to Welsh Office Circular 53/88 regarding the use of the Welsh language as a material consideration in development control and development plans (Welsh Office, 1988). The Circular, which is being replaced by a Technical Advice Note, requires local authorities to recognise the Welsh language as a planning issue, even though it is not strictly a 'land use' matter. Indeed, its role as a social issue within the planning arena is still giving cause for concern to local planning authorities over ten years after its introduction, since little guidance has been provided to planners on how to implement the contents of the Circular. Secondary advice from Central Government, through Circulars and Planning Policy Guidance notes, could well be used to deal with second homes where areas are considered to be under-pressure. What would be far more problematic, however, is stating how any future advice could be implemented within the existing statutory planning framework.

Circular 30/86 was used as a *positive* mechanism with which to provide for future housebuilding for particular social groups. Circular 53/88 *positively* makes the Welsh language a planning issue. Department of the Environment Circular 13/96 provides guidance on enabling affordable housing to be recognised within the planning system, again a *positive* move. The problem in covering the second homes issue through the release of a national policy document is that the underlying rationale for intervention in certain areas of the

UK at least appears to be, essentially, negative; to prohibit or scale down the availability of housing for second home purchasers in rural areas. The British planning system is positive, not negative. National planning policy guidance states quite categorically that:

> It is not the function of the system to interfere with or inhibit competition between users of and investors in land or to regulate development for other than land use planning reasons. Applications for development should be allowed... (Welsh Office, 1996b, para.7).

The implications of this policy requirement are that it is virtually impossible to act negatively *per se* to prohibit future use. Under the present arrangements, a policy Circular could be introduced to *promote* further second homes, but not vice versa. If, however, the planning system operated negatively, that is, there was a presumption *against* new development unless there were strong contrary reasons, it would be possible to introduce advice that restricted future development. This latter change, which would involve up-turning the entire land use process, would require the release of a national planning policy document; there would be no need to introduce new legislation, since the current presumption in favour of development is merely a policy, as opposed to a legal requirement. Some political parties have toyed with this idea over the years (on the basis that it would promote the environment as the primary consideration) but have since ruled out the change due to impracticalities.

How, then, could limiting second homes be acknowledged within the present planning system? It appears that the only way forward is to base the *arguments against* on strictly planning criteria, and then to present the problem as one existing at a local level. This would necessitate proving, beyond all reasonable doubt, that the existence and supply of local affordable housing was an interest of acknowledged importance that would be demonstrably harmed by the provision of further new housing that might be within a high price bracket. Within socio-culturally sensitive areas, such as Welsh-speaking rural Wales, a second and related interest of acknowledged importance could be the Welsh language, since provision already exists through the advice contained in Circular 53/88. If a local planning authority was considering using these issues to overturn the presumption in favour of development at the local level then a substantial amount of evidence would be required, firstly relating to language erosion (and related service disruption) and secondly, relating to the scale of the housing shortage in localities where a planning response might be warranted. The policy relating to the Welsh language, has not, as yet, undergone any testing in the British courts. In terms of affordable housing, policies to generate future housing could be employed but the control of second home purchasing could not be directly dealt with; control over existing stock could, however, be subject to a locals-only restriction if houses were owned by the local authorities. Indeed, a future Government that proposed the reintroduction of council house-building (and additional grant support for the voluntary sector) could use the locals-only restriction and local covenants as an effective means of controlling the influx of second or multiple-home purchasers into rural settlements and also retain an appropriate stock of affordable housing for local people. Needless to say, such a policy change would require strong backing and commitment at a national level and would, in any case, operate outside the planning system, requiring amendment to the 'right-to-buy' legislation.

Planning Controls in Context

Problems created by second home ownership in the effective supply of rural housing stock would warrant new legislation at the national level, and change of use control may offer a solution. However, it will not be without its drawbacks and these must be carefully considered before new measures are implemented. An important principle to bear in mind is the fact that planning controls will form just one element of a wider policy approach required to manage second homes and rural housing problems more generally. They must be complemented with continuing support for the housing functions of the public and voluntary sectors and the recognition that peoples' leisure aspirations have to be accommodated across both urban and rural landscapes. The planning system has a positive role to play in assisting the provision of local affordable housing. The lack of affordable housing in rural areas could well be an underlying reason for both a shortage of housing and for the stock that exists to be gradually withered-away through house purchasing by 'non-local' people. If local authority planning policies and decisions can now legitimately deal with the shortfall of local affordable housing, the increase in new affordable stock could mitigate against the numbers of existing stock becoming second homes. This might go some way towards indirectly controlling second homes through planning, although it should not be used as a direct means of inhibiting second home development *per se*.

Measures which seek to both provide local affordable housing and satisfy the demand for rural second homes in a locality must ensure that the economic benefits of development are maximised while social costs are minimised. Second homes do possess associated costs *and* benefits. This book has attempted to show that the costs are most readily associated with second homes in the existing housing stock. In contrast, benefits may be derived (with the assistance of appropriate planning controls in other areas) from purpose-built second homes. By adopting a strategy of control and development, the aspirations of second home seekers and the needs of rural populations may be successfully balanced. However, there is a need for considerable more research in a number of areas.

First, it must be realised that policy 'packages' will be tailor-made for specific areas; it is not possible to develop policy recommendations for local rural areas without field research. Local research must seek to separate local housing needs from second home aspirations (that is, wider market demand) and consideration should be given to the operation of the housing market, the needs of competing groups and the stock balance. This would include the current second home situation through an examination of the types of dwellings used for this purpose. On the basis of such local research, policies could differentiate between settlement types and associated problems. This provides a general framework but could only be applied given a thorough understanding of the local housing market and local rural problems.

A second area of research might be the role second home owners play in the market for surplus rural stock. After the rules governing improvement grants were amended in 1974 and second home owners were effectively barred from receiving grants, demand shifted to the effective housing stock. Was the saturation of the surplus rural market or the removal of grant rights the root-cause of the cross-competition for mainstream housing which developed after the economic recession of the mid 1970s? Controversially, is there scope for giving second home seekers grants to renovate some rural properties (unclaimed by the local population) where this will have a proven environmental benefit and decrease the competition for mainstream housing?

Thirdly, the potential impact of reform of the planning system needs to be carefully reviewed with particular reference to the additional power that might be. given to local authorities under the delegated arrangements apparent under devolution. Policy and procedural changes to the planning system are on-going and this report has focused on the current system on a UK basis, although some attention has been made to alternative strategies. A future Government in Westminster or in Edinburgh may legally reform the planning system nationally or alternatively elected assemblies may amend the planning policy framework differently in England, Scotland or Wales, associated with moves - particularly at the European Union level - to re-interpret the planning system as a broader spatial planning process involving economic, environmental and social considerations.

9   Local Rural Change

Chapter Six provided an examination of how certain European states have coped with the second homes phenomenon over the last 30 years. This indicated the scale of the issue across the breadth of the European Union but also illustrated the complexity of searching for an appropriate European-wide "solution" to second home growth. Chapters Seven and Eight translated the issue to Britain, by considering both the housing and planning policy options currently available and the potential offered in the years ahead to implement a set of solutions particular to the UK.

This Chapter considers second home growth from the perspective of local rural areas. As has been noted so far in this book, second homes should be viewed within the wider context of local rural change, which could result from not only developments in the countryside and agriculture, but also influences from urban areas and the urban population. Urban changes affect the economic, physical and cultural aspects of rural areas often in ways that stem from outside the nearest urban locations. Serious detrimental impact on rural lifestyles can be caused by, amongst other things, large-scale out-migration particularly of young people, agricultural decline, a narrowing of local economic bases, immigration by older generation 'urbanites', service withdrawal and isolation through peripherality, a lack of housing opportunities for local people, and linguistic changes. Overlying these problems, there continues to exist the romanticised popular perception of rural life.

We would like to turn our attention in this Chapter to the nature of rural change at the local level, by examining five European regions. All five regions, West Sweden, Savo in Finland, Galicia in Northern Spain, the Scottish Highlands, and Gwynedd in North Wales, have experienced rapid second home growth over the last 30 years but each also illustrates the type of rural problems highlighted above. We present each case study analysis with a description of the region by identifying the geographical, economic and environmental context to the rural areas, before going on to examine housing issues generally and second home growth specifically. The case

131

studies are intended to act merely as illustrations of typical rural areas experiencing second home growth Following discussion of the five case studies, we present some overarching conclusions to the nature of local rural change.

West Sweden

Five areas were examined on the Swedish West Coast: Tjorn; Sotenas; Arjang; Dals-Ed; and Tanum. Tjorn is the sixth largest island of Sweden, located approximately 50 kilometres from Gothenburg and 167 square kilometres in size. 14,000 people live on the island which is generally hilly with steep cliffs and fertile valleys stretching inland. Farming, fishing and forestry were the principal industries of Tjorn and although now in decline still generate an immense impact upon the social and cultural life of the island. By its close proximity to Gothenburg and high environmental standards, Tjorn has become an attractive area for urbanites, who have started to purchase properties and commute to the city. Tourism is also on the increase as a result of Tjorn's "quaintness" of old houses, small-scale communities, and fishing villages. Over 45 per cent of the current population has moved in to Tjorn in the last ten years and over one in two of the houses are second or holiday homes.

Sotenas, at 140 square kilometres, is one of the smallest authorities in Sweden with 10,000 inhabitants, located on the Skagerrach coast. Once again, the area has a long fishing tradition although the area's housing is generally concentrated in small towns and villages, most of which have been preserved as old fishing villages that have attracted tourists in increasing numbers. Commuting is on the increase following significant in-migration to Sotenas by middle-class urbanites while out-migration amongst unemployed young people and new university students has also increased. Over 4,300 people have moved to Sotenas in the last ten years (5 per cent of these from outside the country) with just over 3,200 people leaving, while a third of the total population do not originate from the locality. There were 6,697 registered holiday/second homes in Sotenas in 1994 and when compared to the total population figure of 10,000, it is easy to see why second home growth in this region is viewed with concern. The local economy is strong at the present time and the 40 minute commuting distance to the closest town is particularly attractive.

The third area, Arjang, is located approximately 200 kilometres from Gothenburg on the Norwegian border. Population, at 10,000, is sparse and is dispersed across 1,417 square kilometres. The local economy is centred around manufacturing industry, forestry and farming, with growth in electronics, wood-refining and tourism on the increase. The area is environmentally sensitive with an abundance of non-polluted lakes, free camping, lakeside parks and woodlands, and small-scale outdoor recreational pursuits. Significant rationalisation in both the forestry and farming industries have reduced the number of small-scale communities of local people, and the vacant housing has been purchased by second home owners. These incomers have tended to be from outside Sweden and they have brought with them urban attitudes and values. Local friction has resulted between the indigenous population and the newcomers, the latter attempting to protected the area for their own use and failing to recognise the Swedish laws permitting free access and the 'right to roam'. Newcomers total approximately 4,000 over the last ten years with 3700 residents out-migrating. Latest figures indicate the presence of almost 4,000 holiday or second homes in Arjang. At almost 90 minutes from the nearest large town, house prices are lower in this area than other Swedish second home areas although the number of holiday units presently remains stable.

Dals Ed contains just 5,400 people across 728 square kilometres and is located on the Swedish-Norwegian border. Forestry and farming dominate the local economy and community, although tourism in both summer and winter months has started to increase lately since the area is one of the most beautiful in Sweden. The decline in the principal industries has not been offset by the creation of new small enterprises. Well over 55 per cent of the population has moved to the area over the last ten years, displacing a similar number that has out-migrated. Houses have started to be purchased as holiday homes in increasing numbers lately.

Tanum is located midway between Gothenburg and Oslo in Norway, along Sweden's west coast. The area is approximately 900 square kilometres with a population of over 12,000. Farming, forestry and fishing have been the main industries although, once again, tourism is a major employer too. The community comprises densely populated areas although the lack of any large-scale industry has resulted in the region becoming attractive to commuters. The coastal areas are the most sought after locations with small-scale farming, old fishing villages, ancient monuments and environmentally-sensitive locations. There are over 8,600 holiday or second homes; because of the dormitory nature of the settlement, the area is comparatively deserted during the week.

Savo in Finland

Three areas were examined in Finland: Iisalmi; Karttula; and Suonenjoki. Iisalmi is located approximately 80 kilometres from Kuopio on the main highway northwards. It is the administrative centre of the Upper Savo area with over 24,000 of the total 70,000 inhabitants located in Iisalmi. Growth continues despite concern at the future of the local economy, which is centred on agriculture and dairy production. Some unemployment has already occurred as a result of re-structuring with the relatively small-scale industrial and service sectors unable to compensate in employment opportunities. West of Iisalmi is the village of Runni with a population of just over 550 which has been particularly hit by the rural decline. In response, villagers and the Iisalmi city authority have implemented a development project, centred on tourism. With an 18th century health spa in the village, Runni has been designated as a cultural protection area by the Ministry of the Environment and the National Board of Antiquities and Historical Monuments. With funding available to support tourist-related projects, it is intended to protect and renovate the older buildings as an historical spa resort. Over 7,700 people have moved to Iisalmi in the last ten years and the number of second or holiday homes total 1,147 (1997 figures).

Karttula is situated 50 kilometres to the west of Kuopio. Land prices in this area is considerably cheaper than the urban centre, with lower construction costs and taxation. As a consequence, large numbers of people from Kuopio have moved in to Karttula, although the municipality has welcomed the in-migration for the benefit of the local economy and accordingly started to develop further land for residential and commercial purposes. Karttula initiated the Pihkainmaki project in the 1980s to attract new families to the area in a planned town of over 200 houses; this was intended to create a balanced population structure to support the provision of local services and the project has been regarded as a success. Most of the in-migration occurred at this time with a levelling out in the 1990s; Karttula's population was approximately 3,500 in 1995. 1,900 people had moved in to the area over the last ten years. The new residents have been attracted by the coastline location and in the summer months, the local population is doubled in number. Recent figures indicate the existence of over 1,200 holiday or second home units in the area. The municipality's enthusiasm to cater for the in-migrants has led to local problems, however. The distribution of infrastructure costs and public services for new housing developments have been contested by local people, particularly as second home purchasing seems to have already peaked.

The third area, Suonenjoki, is situated 50 kilometres west of Kuopio on the main railway link between Iisalmi and Helsinki. With a population of over 9,000, the area is declining as a result of local out-migration. Holiday home purchasing was particularly high in the 1970s with the municipality developing the shoreline to benefit housing development, mainly to benefit the local unemployment rates caused by industrial restructuring. Second home numbers currently stand at 950 with around 2,500 people moving into the area over the last ten years.

Galicia in Northern Spain

The Ribeira Sacra Lucense area of Galicia is a beautiful rural farming area with great tourist potential. The local economy is weak based on the existing primary industries and there is high unemployment and large-scale out-migration. Municipality officials intend to revitalise the area with investment based mainly on tourism and complementary industries, capitalising on some important historical centres and the natural environment. Sizes of the settlements in the area range from 3,200 people (in Pobra de Brollon) to 10,500 (in Chantada). Most areas suffered high numbers of out-migration, particularly of younger people, in the 1960s and 1970s as a consequence of the steep decline in agriculture. Even today, the percentage of young people under the age of 14 is approximately 12 per cent, compared to a third of the inhabitants who are aged 65 or over. In some areas, this is of particular concern. In Pobra de Brollon, for example, only 9 per cent of the population is in the 0-14 age bracket, compared to 27 per cent between 45-64 years, and 33 per cent over 65 years. Some of the population who emigrated in the 1960s and 1970s are now starting to return to retirement homes in the region; this is most noticeable in Carballedo and Savina. This retired in-migration has also been accompanied in the 1990s by increased mobility within and between rural villages within the region.

Agricultural production remains in evident, although in some areas dairy and cattle farming are in decline. In some locations, some industry is present in the form of workshops, textile and footwear production, and vineyards, although all these remain relatively small-scale. Unemployment rates vary between 7 per cent in Carballedo to 19 per cent in Monforte. Major transport connections are good with low journey times to the larger neighbouring towns and cities of over 50,000 people. In addition to the rail link through the region, the economy has been aided by the construction of a dual-carriageway highway crossing the area from west to east with good links between settlements. The region is culturally and linguistically sensitive, with Galician spoken and taught in local schools in addition to Spanish. The area is environmentally-sensitive and there is currently a project between the Galician Government and the European Union to declare the Ribeira Sacra and River Sil Canyon as Natural Parks with accompanying environmental protection measures.

Land prices in the region are low mainly due to the traditional pattern of small farm-holdings although prices in the urbanised areas and on the urban periphery are starting to rise sharply as a result of urban in-migration and retirement plans of people returning to the area. House building is occurring in these locations but on individual plots of land. More formalised house building through planning mechanisms is not apparent due to the lack of a statutory planning framework. Second home numbers vary between settlements although are present in all locations. In 1993, there were 92 in Carballedo, 235 in Pobra de Brollon, 399 in Savina, 628 in Monforte, and 642 in Chantada. Second home growth is not viewed with particular concern in Galicia; more worryingly is the unbalanced age structure in some areas and the high numbers of derelict properties, that mostly outnumber second homes.

The Scottish Highlands

Four areas in the Scottish Highlands and Islands were identified for analysis: Strathspey villages; Ardnamurchan Peninsula; North Mainland of Shetland, Yell and Unst; and Sleat Peninsula. Each of the areas are characterised by the sparse population, remoteness and varying local economic bases, in addition to a proliferation of second home purchasing.

Strathspey villages are located in the central Highlands, immediately to the west of the Cairngorm mountain range. Comprising Aviemore, Carrbridge, Boat of Garten, and Nethy Bridge, the villages have attracted a significant number of in-migrants over the last 15 years as a consequence of the beautiful scenery, the villages' proximity to the snow resorts of the Cairngorms, and the availability of holiday homes, and the population over this time has increased. Most of the in-migrants are young people or families, attracted to the area from the Scottish "central belt" of the cities and from England for the outdoor pursuits, environment and skiing. Some of the holiday home accommodation has been purpose-built, through "timeshare" apartments, for example. But the majority of second homes have been permanent residences from the existing housing stock. In some locations, the proportion of second homes amongst the existing housing stock is around 20 per cent. In 1997, it was estimated that there were 292 second homes and 300 holiday homes in the area, with 124 empty properties. The external demand for second homes has raised property prices in the area significantly over the last ten years to levels beyond those of neighbouring areas and the affordability range of many local people. The planning system has also affected prices because of the high conservation value of the environment and the strict regulatory policies in place prohibiting development.

The Ardnamurchan Peninsula is a remote rural area situated on the west coast of Scotland. The population of 380 comprises dispersed settlements of crofting townships linked by single track roads, with the nearest town, Fort William, some 40 miles away. There is a heavy reliance on agriculture to sustain the local economy and also on fishing. The landscape is extremely attractive and there has been a steady and growing demand for second homes and holiday homes, mainly by young people and families possibly linked to the recent tourism developments. Because of the scenery and geographical location of the settlements, tourism is on the increase and has become essential for the local economy although the area's peripherality means that it is some distance away from the major tourist networks. The resident population is predominantly retired and declining in numbers. As a result of the decline in the local economy, thanks to agricultural stagnation, the area is depressed and there has been little new investment over the last few years with the result that services are scarce; the school disappeared some time ago. The lack of services have had a serious impact on the attractiveness of the area to stimulate the local economy. Those houses that do come onto the market are usually bought up by outsiders at extremely high rates for use as holiday homes. In 1997, there were 114 second or holiday homes on Ardnamurchan The local cultural traditions associated with the crofting communities are perceived to be under threat as a result and local authority officers are concerned as to how to stop further decline. 30 per cent of the people speak Gaelic.

The North Mainland of Shetland, Yell and Unst, with a population of approximately 4,500, is amongst an island group located beyond the mainland of Scotland. The main centre of administration, employment and service provision is in Lerwick, in the south Mainland which houses approximately one third of the population. In-migration has occurred due to the presence of both oil-related activity and military establishments and young adults and families figure prominently. The local economy has been in generally good health throughout the 1980s and 1990s thanks to the nearby proximity of the North Sea oil field which dominates but tourism has also contributed. Unemployment levels are therefore comparatively low, although the north Mainland's economy remains uncertain at the present time as a consequence of its reliance on crofting, agriculture, quarrying, fishing and textiles. Culturally, the island group is very different to the rest of Scotland, resting on Nordic traditions. With continued uncertainty over the future of North Sea oil and military bases in the medium term, alternative economic bases are being sought to provide future stability. 40 second homes and 36 holiday homes were noted in 1997 although a further 222 were empty properties.

The housing market is not particularly buoyant principally due to the inaccessible location of the islands, the construction costs associated with new house building, and low salary levels of the local population. Identifying new housing land is problematic because of the traditional patterns of land tenure.

The Sleat Peninsula is located at the most southern point on the Isle of Skye, western Scotland. Despite its peripheral location, the area is experiencing growing population levels with increased interest in the housing market and recent infrastructure improvements, tourism developments, education establishments and hotel and conference facilities, that have opened the area up to economically active outsiders. Attempts to expand these types of development recently have been met with opposition on environmental and socio-linguistic grounds from the local population; well over half the population speak Gaelic. The number of second and holiday homes - over 100 on Sleat - is high as a result of the area's high quality environment, and house prices have grown significantly above the national average over the last ten years. Although in-migration is evident within young people and families, one fifth of migrants to Sleat are retired. The social rented sector is pressurised and despite the healthy economy, homelessness and the lack of local affordable housing is a continued problem, many households have resorted to residing in caravans and other forms of temporary accommodation throughout the year.

Gwynedd in North Wales

The North Wales area, centred on the Lleyn peninsula, is traditionally referred to as "Welsh Wales", since it hosts a high proportion of Welsh speakers and contains some of Wales' finest scenery around the coastline and the Snowdonia mountain range. There has long been concern at the influx of second and holiday home owners in the area, accompanied by the decline in rural areas and concern over the future of the Welsh language. Four areas were examined: Blaenau Ffestiniog; Aberdaron; Llanengan; and Pwllheli.

Blaenau Ffestiniog, a town of almost 5,500 people, is the centre of the former slate quarrying area and is now suffering from the impact of industrial restructuring with low employment opportunities. As a consequence of the lack of sufficient numbers of alternative employment, many young people have out-migrated from the area. This has impacted upon the cultural and linguistic profile of the area. No new housing has been built in the town for almost 100 years and recently the town has become home to families from so-called "problem" housing estates from urban areas across the border in England. This influx of English-speaking people has not been particularly welcome by the Welsh-speaking indigenous population. There are approximately 200 second or holiday homes in the town.

Aberdaron is centred in attractive landscapes and small rural settlements with a sparse population. The area possesses a strong sense of community and is well known for its rich Welsh culture. High levels of out-migration among young people have been experienced as a result of the decline in Welsh agriculture; the population dropped significantly between 1981 and 1991. This figure has not been offset by the in-migration of retired elderly people in large numbers. Tourism and agriculture are the main sources of economic activity although this is seasonally dependent. The housing market is at a low ebb with the low economic base resulting in poor condition of the housing stock.

Llanengan, a community of 800 people, has experienced major development as a tourist centre but has also witnessed significant population restructuring. The main town in the area, Abersoch, is located on the Lleyn peninsula coastline and boasts a boating centre and attractive harbour. With tourism as the main economic activity, out-migration has been high among young people but large numbers of wealthy and retired people have in-migrated. Over 55 per cent of in-migrants have originated outside of Wales in the last ten years. House prices are characteristically high and only affordable to the in-migrant population. Llanengan community itself has the highest proportion of second homes in North Wales with some 400 units.

The predominantly urban area of Pwllheli, the main service centre of the Lleyn and the town with the highest proportion of Welsh speakers in Wales (80 per cent), is set within a rural area whose population structure and profile has remained relatively stable. The economic base is quite strong with service and manufacturing employment, particularly related to marine activities. Out-migration has only been witnessed from qualified younger people seeking better employment opportunities elsewhere. House prices are comparatively cheap compared to the prices within the surrounding rural area. Although popular with visitors during the tourist season, Pwllheli has not been noticeably affected by second home growth.

Key Features

The extent of second home growth in our case study areas is extremely variable, reflecting the differences in the extent of rural change being experienced, the economic conditions prevalent locally, and the role and purpose of the communities in employment and tourism development. Regions in each of the five areas are all experiencing serious agricultural deprivation and decline of rural services as a consequence of a narrowing of economic bases and opportunities. But each of the five areas also appears to be at different stages of rural decline. In many of the areas, out-migration of young people and in-migration by retired or semi-retired urbanites has transfigured rural communities into becoming centres of "residential tourism" (Gallent, et al, 1998). Where this change has occurred in areas of cultural and linguistic sensitivity, the impact of second home purchasing has been particularly noticeable.

In the Scottish Highlands, rural change has been most pronounced as a result of agricultural decline, population loss, the unique forms of land ownership and tenure, reduced opportunities in the housing market, and the peripheral location of settlements. West Sweden and North Wales are also experiencing similar problems with a narrowing of the local economy and an inability to foster new forms of economic activity other than through tourism and the concomitant growth of temporary and permanent in-migrants. In all three areas, economic depression and reduced housing choices for local people is deepened further by a perceived "threat" culturally and linguistically. Attitudes amongst local people to these changes can be quite strong as they witness the selective out-migration of their most educated and gifted young people to seek employment and housing opportunities elsewhere. Nevertheless, it would be wrong to focus in on second home growth as the sole cause of these changes, and in some areas the growth in second home purchasing is not a root cause of change but merely an externality caused by wider rural restructuring.

Different regions of the European Union view this cause or effect symptom differently. In Galicia, for example, second home growth is not viewed with the same degree of concern as attitudes in North Wales and the Scottish Highlands. Local municipality administrators in Northern Spain recognise the contribution provided by second and holiday home purchasers toward the rehabilitation of derelict properties. In Savo, council policy-makers and politicians implemented a series of measures in the 1970s and 1980s to attract second and holiday home purchasers to the area for local economic purposes, including the provision of infrastructure that locals could benefit from too. And in all areas, the tourism benefit offered by second and holiday home owners is recognised. The problem stems from the low choices available in the housing market for local people in rural areas that are restructuring towards a tourism reliance.

A narrowing of the local economic base principally in agriculture, but in some areas additionally from other forms of industrial decline (for example, slate quarrying in North Wales, textile decline in Northern Spain), has resulted in communities and officials searching for ways in which the economy can be rejuvenated. There have been little opportunities locally to stimulate economic development and regions and communities have sought to rely on external sources of revenue to generate alternative sources of employment. In all five case study areas, this alternative source of local economic revenue has been provided by concentrating efforts on tourism. The picturesque landscapes in each location, the environmentally-sensitive areas, the deep rural profile of communities and the remoteness or peripherality of communities from the main urban centres, have been the attractive features for urbanites searching for the elusive "rural retreat". Regional and local officials have exploited urbanite concepts of the rural idyll, directly or indirectly, by promoting policies and projects that potentially promote tourism in their areas as a way of rejuvenating local economies that are otherwise spiralling downwards thanks to agricultural decline. Resources have also been available too through finance from regional and national governments, and the European Commission, to fund the costs of tourism developments in these rural areas. With little in the way of alternative economic development, there can be little blame attached to officials and politicians in their desire to stimulate local economies and create some wealth for rural communities. But a consequence of their actions has been in most cases to contribute towards in-migration by appealing to urbanite perceptions of picturesque rural lifestyles. With the growth in tourism in these remote rural locations, second home and holiday home purchasing has also increased, which has only created pressure on the existing housing stock, on housing options for local people, and on the population balance of communities. The tourism solution to local rural economic decline, therefore, has often resulted in a deepening in the rural problem in different ways, socially, culturally and linguistically.

The social restructuring of rural areas caused by in-migration of retired and wealthy commuting urban people, associated with the loss of young people out-migrating, has led to an imbalance in the age profile of areas, a fact most noticeably observed in Northern Spain but also prevalent in Wales. The in-migration by these sections of the population also stems from the attractiveness of the rural idyll. Out-migration amongst young people is directly related to the lower opportunities available in the remote rural locations in anything other than low-skilled or tourist employment. In Scotland, however, this trend was reversed. In the remote rural areas on Shetland, for example, there has been an influx of young people and families thanks to the close proximity of economic activity associated with the North Sea oil field and Ministry of Defence establishments. In the Strathspey villages, too, young people have moved in attracted by the environment, the snow resorts and wilderness of the Cairngorms. In both areas, this influx has originated from outside the area, either from urban locations in southern Scotland or else from England. This certainly brings into question whether sustaining large numbers of young people for economic purposes in these remote rural areas would in fact lead to a rejuvenation in the fortunes of the communities. In both examples, second home and holiday home growth has still been viewed as one of the causes of rural community decline. We therefore conclude that although a balanced age profile among the population of these rural areas is indeed desirable, it is only one aspect of rural community profiles. If a more balanced rural community can be achieved, there still remain issues and concerns relating to the attractiveness of remote rural idylls among older generation urbanites from elsewhere, the requirement to foster economic activity through tourism, and the problems in securing affordable housing for indigenous low income groups.

We should also remember that differences occur within regions, in relation to the economic fortunes of remote rural areas. In both Northern Spain and in North Wales, for example, there were differences between settlements in economic terms and in the perception of the existence of "a second home problem". This is possibly related to the economic cycle of areas, the demand for second and holiday homes at any given time, the personal disposable incomes of new residents, and the attractiveness of second home living in selectively-attractive locations. Afterall, in many of these regions, and indeed in the examples illustrated throughout this book, second home growth has been a feature of the last 30 years. Perceptions and attitudes differ constantly and the desirability of a remote rural area may not last more than a few years once an area becomes "too popular". As a consequence, the search for a particular policy or political solution to second home growth in any community is extremely difficult, not only in relation to the various socio-economic, environmental and cultural factors present and in varying degrees in selective locations, but also in assessing the likelihood of the problem continuing to exist for more than a few seasons.

Perhaps the answer lies in developing community-based mechanisms to foster "bottom-up" approaches to stem rural decline, rather than impose "top-down" solutions, and thereby contribute towards a revitalisation of community development based on sustainability and socio-cultural values. This is, of course, no mean feat, and we recognise that community development can only be achieved through the active participation of a variety of official and voluntary agencies, in addition to the active support of both the indigenous population and migrants. But cultural, environmental and economic diversity within these remote rural locations can at least be expressed more formally to the representatives of the policy-making agencies. It may also be an appropriate method for harnessing the cross-cutting problems and perceptions associated with rural second home regions.

The selective case illustrations discussed in this Chapter, together with the European review undertaken in Chapter Six, suggest that rural regions experiencing second home growth are at various stages of prevalence. As a result, there seems great potential for these regions to learn a great deal from each other and to foster transnational co-operation on ideas for community capacity-building. This should not be viewed as attempts to impose a European-wide solution, but rather a useful way of exchanging ideas, concepts and policy initiatives. The next Chapter considers some of these ideas in further detail by addressing ways forward in research and policy development.

10 Conclusions

Introduction

The expansion of second home ownership in the United Kingdom and across Europe is indicative of social restructuring and rural change. However, second homes are not only symptomatic of change, but may be viewed as a component of that change given their position in the housing market and their relationship with other components and the macro-economy. We conclude our review by returning to the principal themes addressed throughout the book.

Concepts and Information

Conventional wisdom dictates that 'second homes' are 'occasional residences' and used primarily for 'recreation purposes'. The user often resides some distance from his or her second home (depending on transport considerations and the frequency of use) and is normally a free-holder. A distinction is often made between 'permanent buildings' and static caravans (Shucksmith, 1983). Built second homes in the countryside have been the subject of this report.

In the UK, it is difficult to provide an accurate statistical portrait of the number of second homes given the lack of comprehensive data. In 1985, there were estimated to be 221,000 built second homes in England and Wales (Hansard, 1985), a figure which is likely to have risen to a quarter of a million today.

Empirical studies have outlined a number of data sources used in the study of the second home phenomenon. These range from national population censuses to local rating and electoral registers. Other sources may include tourism and economic development reports or housing and planning surveys. National sources are often deficient either because they underestimate the extent of ownership (as is the case with the British

Census) or are susceptible to false registration (as in Austria). Local surveys are motivated by local issues and nationally, provide a fragmented picture. A lack of consensus over definition also militates against accurate national or international assessment.

The Origins of Growth

Structurally, growth is driven by regional economic inequality. In 'exporting regions' (that is, those areas where demand originates), greater affluence and more leisure time is linked to the potential demand for second homes. In the 'importing regions' (peripheral areas where demand is satisfied), less affluence and economic stagnation fuels depopulation, increasing the availability of surplus housing stock (which is a finite resource and diminishes over time), often the focus of second home demand. The differential between property prices in the exporting and importing regions influences middle class housing choices, creating a unilinear demand flow. These structural determinants combine with a 'cult of nostalgia' for the countryside (Newby, 1980b) and a relative lack of recreational outlets in many urban environments to produce specific 'leisure choices'.

These choices are influenced by personal motivations (turning demand potential into effective demand) which may include the conferring of status, access to specific rural resources, empathy for a particular locality or a host of other factors, too numerous and varied to catalogue. Robertson (1977) has suggested that decision making is based around 'utility evaluation'; the decision to acquire a second home is dependant on the notion of 'anticipated utility', whilst subsequent decisions concerning the acquisition are based upon 'actual' and 'projected' utility value.

The growth in second home ownership has been characterised by a process of 'social democratisation' whereby owners are increasingly drawn from less affluent socio-economic groups. Whilst some commentators have welcomed this 'extension of privilege', others have underlined the heightened social tensions which may be generated in some areas and compared this phenomenon to the widening 'privilege gap' between exporting and importing regions.

146

Ownership and Demand

The spatial distribution of second homes within importing regions is determined by a number of factors. These might include the distance from major population centres, landscape characteristics, particular physical features, recreational resources and services and the availability of suitable property. Improved road networks across Europe are opening up new areas to second home seekers whilst the established second home areas are sustained by a process of colonisation whereby existing owners promote the area to friends and relatives. The abandonment of rural dwellings by the host population has been a key factor in determining the distribution of demand. An equally important process in France has been inheritance with younger urban generations inheriting rural property from parents of grandparents. Distribution, like demand, is dependent on a number of structural and personal factors and is therefore often difficult to explain in simple terms.

Emphasis in this book has been placed on the different types of dwellings used as second homes. A typology is identified which includes those dwellings which are surplus to local needs (often empty or derelict), dwellings drawn from the mainstream housing stock and purpose-built second homes. It is necessary to distinguish these types in order to consider their relative impacts and to formulate policy. It is certainly wrong to assume that all second homes are alike.

Different continuums of types might be identified in different regions. In England and Wales, up until the early 1970s, demand was focused in the surplus housing stock. After the economic recession of the mid 1970s, the rapid depletion of the surplus rural dwellings and the removal of improvement grant rights for second home owners in 1974, demand shifted to mainstream housing. In general, demand seems to have remained in this part of the market. Restrictive planning controls (compared to many of Britain's European neighbours) appear to have prevented some of this demand from being diverted to a relatively insignificant purpose-built sector. The impact that second homes have on the housing market and the social disruption (if any) they cause will be dependant on the types of dwellings used for this purpose in given areas.

An extrapolation of trends, based on the past and current distribution of second homes, will be essential in the formulation of policy, either nationally or locally. However, past analyses of the growth phenomenon tended to provide inadequate predictive models, mainly because they focused on processes in the importing region rather than adopting a more holistic approach.

Economic Costs and Benefits

The impacts of second homes are invariably subject to local circumstances and can only be understood in the context of the local housing market. However, some generalisations can be drawn.

House prices may be adversely affected where demand for second homes is concentrated in the mainstream housing stock; house price inflation may lead to the exclusion of lower income locals from the housing market (either because they are unable to compete with relatively affluent incomers or because of a decline in rented accommodation). However, local house prices are also a function of the preferences of higher-income locals, restrictive planning measures and the overall adequacy of local housing supply.

The acquisition of second homes may bring some economic benefits; these will be most significant where the property acquired was outside the effective housing stock. Acquisition of mainstream housing may accentuate housing shortages, although the supply of rented accommodation is more likely to have been affected by successive Rent Act legislation and the right-to-buy than the arrival of second home owners.

Property speculation in the second home market was an important issue in England and Wales in the early 1970s although this has now been suppressed by the removal of improvement grant rights.

General housing stock improvements will accrue when demand is focused in the market for surplus dwellings (as was the case in France when British buyers arrived in the late 1980s). Local contractors may benefit from renovation work whilst additional income and employment boosts the local economy. However, where demand is focused in the effective stock, improvements may not be necessary and any economic benefit is likely to be off-set by additional social costs.

The general economic contribution of second home owners (during their visits to the importing region) is extremely difficult to assess and there are few empirical studies. However, where second home owners displace local residents, it is likely that their contribution will have been less than the potential contribution from the displaced resident. On the other hand, where owners occupy dwellings outside the effective stock, the economic contribution of these incomers is likely to be substantial in some local areas although the effect on the national or regional economy may be slight. In terms of Council Tax contribution, the arguments outlined above will apply equally. A confusing factor, however, is that the contribution of visitors in isolated dwellings is likely to be insufficient to pay for additional social services if they eventually retire to the area. In fact, elderly retired persons are likely to place significant strains on the local economy and local tax payers in some areas. However, past research suggests that fewer second home owners are likely to retire to the importing region than in-migrant non-owners, mainly because they have a greater awareness of the practical problems associated with rural lifestyles.

The economic contribution of second homes cannot be assessed without a closer examination of local housing markets as it is dependent on the degree of separation between first and second home demand. Where there is direct market competition and the exclusion of locals, any economic contribution is negated by the loss of locally-derived revenue and adverse social impact. However, where there is clear separation between the markets (as is the case in France and undoubtedly in certain areas of England and Wales), the economic contribution may be locally significant.

Environmental and Social Impacts

The environmental impact of second homes is again dependent on the types of dwellings used for this recreational purpose. Generally, impacts in this category relate to the general growth in leisure activity in the importing region (of which second homes may, or may not, be a significant component), increases in development pressure where the market is dominated by purpose-built second homes (commercial holiday villages fall outside this category), and the landscape and environmental impacts brought by the restoration of surplus rural housing.

The reviewed literature suggests that the landscape benefits of restoration are substantial and the conversion of some surplus dwellings may lead to the preservation of buildings which are of historic or architectural interest. The impact of new-build will be dependent on the stringency of planning control although it has been suggested that where design and siting are carefully regulated, the environmental impact of new developments range from being 'neutral' to be being beneficial.

An analysis of social impact must be grounded in an understanding of competitive housing markets and their operation in contrasting local contexts. Second homes may have an adverse social impact where incomers compete directly against locals for mainstream housing, although inflationary pressure on house prices is principally a function of market supply and the way in which planning influences this supply in certain areas. Where social tensions are most apparent, local authorities may seek to control the occupancy and supply of new housing, further fuelling inflationary pressure and exacerbating social tension.

However, in the conventional model, rising house prices lead to the displacement of permanent residents and the disruption of local services as regular demand declines. As incomers take over existing dwellings, those locals which remain occupy non-market housing and this cleavage comes to symbolise the way in which rural markets have been restructured.

Attitudes have been seen as an important means of gauging social impact. Bollom (1978) has argued that those communities least affected by second homes (but which anticipate the threat) are most vociferous in protest as local institutions, transformed into opposition structures, remain intact. In contrast, those communities most affected by temporary incomers are characterised by the submergence of local institutions and a lack of protest. Silence, not protest, is the hallmark of acute social disruption.

Other observers argue that the 'loss of community' is more accurately attributed to underlying socio-economic forces. Arguably, rural communities, by becoming less 'traditional', are becoming less recognisable and a fear of change is generating a hostility towards those features of the rural landscape which have come to symbolise that change.

European Perspectives

The purpose of this discussion was to consider how different state systems in Europe have responded to particular second home issues. Various experiences were examined in Denmark, Finland, France, Spain, Sweden, Norway, Greece and Belgium. However, contrasting socio-economic circumstances, the fact that markets are often dominated by foreign buyers and the role of devolved state structures mean that it is extremely difficult to draw comparisons with the situation in the United Kingdom, which is only just commencing devolved forms of policy-making.

For the same reasons, it is unlikely that policy measures implemented in Åland or Sweden would have any practical use in North Wales or in the Lake District for example. The foreign examples, however, discussed in both Chapters Six and Nine, further demonstrate the benefits that second homes can bring in some circumstances (for example, improvements to the rural housing stock in France and Northern Spain) or the problems they may create (or accentuate) where conditions are less favourable (for example, the elevation of house prices and social disruption on the Swedish west coast). They demonstrate that the same divisions in second home types and the same range of related problems occurring in the UK have their parallels abroad.

Housing Policy Responses in the UK

Policy options to regulate second homes may be locally or nationally based, may attempt to enhance housing supply or regulate second home demand, and may fall into the category economic and social development, housing policy, or planning and development control.

Promoting economic and social development at the local and national level is seen as the best long-term mechanism for bringing the housing and employment chances of rural households into parity with their urban counterparts. Once rural households are the competitive equals of urban-dwellers, the housing chances of local residents will increase whilst the housing choices of non-residents will adjust accordingly. However, it is recognised that economic development offers no overnight solution and therefore intermediate steps may need to be taken.

Solutions focus on the supply of rural housing and external demand. Arguably, mechanisms need to be found which offer a means of increasing non-market housing options for local people. These might include the provision of new local authority or housing association dwellings, the 'municipalisation' of empty properties, encouraging housing associations to manage empty privately-owned dwellings or experimenting with community self-help schemes. In recent years, emphasis has been placed on synthesising the housing and planning functions of local authorities in order to procure affordable housing. However, all these mechanisms suffer from the problem of insufficient funding and a lack of commitment to a comprehensive rural housing programme.

The alternative option is to reduce external demand for rural housing. Firstly, this might be achieved by increasing the price of second homes through general taxation, the levying of differential Council Tax (or a standard purchase levy), or restricting mortgage finance. These options would be difficult to implement and could not differentiate between second home types and would therefore undermine the contribution that some second home owners make. Secondly, a property licensing scheme might be used to regulate real estate acquisitions or housing sub-markets could be used to exclude non-local buyers. Such schemes identify the problem associated with second homes located in the mainstream housing market, but would be politically unattainable given the dominance of Britain's market ideology.

Current planning mechanisms are unable to regulate the occupancy of existing housing whilst restrictions on new supply and the exclusion of outside buyers (through the use of occupancy controls) place additional pressure on the existing stock, accentuating price rises and further disadvantaging locals in the market for old housing (and increasing their dependency on new affordable development). This process has led some commentators to advocate means of controlling the demand for second homes in the existing (and mainstream) market. Political weight has been put behind the idea that a planning system, which has paid increasing lip-service to local interests in recent years, should be used to control change of residential use from that of primary to secondary dwelling.

The British Planning Policy Process

Chapter Eight examined the current planning system in Britain and the regulation of dwelling occupancy. It concludes that the control of new housing units tends to focus additional demand on existing housing (accentuating house prices and associated social problems). For this reason, attention shifts to the proposed 'change of use' legislation. However, it is suggested that such legislation may be difficult to implement and could have serious negative side effects. For this reason, other aspects of the planning system are examined and a range of alternative policy options were considered. These include the following issues.

Defining the essence of second home problems: Local authorities should identify the scale (and nature) of the second home phenomenon as existing across the whole of their administrative areas. This would involve an assessment of the use of existing stock, future stock, the dwelling types involved (new, existing and derelict), in addition to the category of settlement within which second homes are located. The present and possible future demand for second homes should also be identified and the effect this would have on the housing market. Assessment should also be made of the local economic benefits accrued from second homes in the local authority area. This would be the first focus of essential research.

Assessing local housing needs: Local authorities should ensure that they have undertaken a comprehensive survey of local housing needs across their administrative areas from which they would be able to develop local affordable housing policies for inclusion within their development plans and use the data for planning control purposes. The results of the surveys would also assist in identifying areas under most pressure from new housing developments.

The inclusion of policies for local affordable housing in development plans: Based on the results of the local needs surveys, policies should cater for any demand for local affordable housing and must be clearly justified with reference to these studies.

Demonstrating the locally important circumstance of culture and language: Areas that are particularly 'socially-sensitive' (that is, those localities that possess a high proportion of minority-language speakers or are culturally significant) should be thoroughly surveyed by local authorities to demonstrate the local importance of the phenomenon. Local authorities may be able to demonstrate locally justified circumstances which outweigh the application of a nationally-formulated policy.

Use Class Order: The introduction of amendments to the Use Class Order 1987 in the UK and the subdivision of the residential category into 'primary residential' and 'secondary residential' functions. The benefits of introducing second home problems into the planning system might be outweighed by adverse effects on the property market, neighbouring authorities and communities, and could lead to permanent socio-cultural reconfiguration.

Occupancy controls: A tightening-up of occupancy controls for new development. Such a move could prove problematic if attempted within the present British planning system but could possibly be achieved through covenants imposed on council and housing association properties. Such moves should be combined with new council housing association provision.

Improvement grants for second homes: The reintroduction of improvement grant eligibility for second home owners could offset the competition and pressure on the mainstream housing stock and thereby 'release' more affordable housing for local people. Associated planning, economic and environmental benefits would include reusing redundant houses that would otherwise remain derelict and/or be of little interest to local people.

A policy presumption against development: Such a radical change would overturn the operation of the land use planning process in the UK at least, but might benefit particular local areas for environmental or social reasons. It is unrealistic to expect a presumption change to occur nationally but it might be enforced within individual sensitive landscape areas (for example, national parks, green belts and areas of outstanding natural beauty) and in areas where there are significant social pressures (for the continued survival of minority languages). A negative presumption would introduce automatic refusals to applications unless the applicant demonstrates the 'community benefits' of the proposal that would befall the area or the minimal environmental degradation associated with the development. It would introduce a 'burden of proof' on the part of applicants. However, any restriction on housing supply may have an adverse effect on existing house prices and a presumption against development will not prevent the change of use of existing dwellings.

In the final analysis, the way forward seems to lie with *increasing housing supply* rather than tinkering with the demand side of the equation; attempts to control occupancy or curb development skirt around the edges of the real issue which is centred on the existing housing stock. Effective legislation to control occupancy in this sector may prove elusive and therefore only by increasing supply may tensions be eased. However, all new house building must be subject to local circumstances and pay particular attention to cultural and linguistic issues which may be especially relevant in certain areas.

Local Rural Change

With regard to wider changes occurring within those rural areas experiencing second home growth, thoughts could turn towards the mechanisms in place within communities to foster partnerships, ideas and projects. Such community-based mechanisms to foster "bottom-up" approaches to stem rural decline, associated with developments at the national level on housing and planning policy, could contribute towards a revitalisation of community development based on sustainability and socio-cultural values. This would also require the active participation of a variety of official and voluntary agencies, in addition to the active support of both the indigenous population and migrants and above all the legitimacy of national and regional legislative and policy-making assemblies to implement the measures. Cultural, environmental and economic diversity within these remote rural locations can then be expressed more formally to the representatives of the policy-making agencies, and would also be an appropriate method for harnessing the cross-cutting problems and perceptions associated with rural second home regions.

Rural regions experiencing second home growth are at various stages of prevalence, and there is a great potential for these regions to learn a great deal from each other and to foster co-operation across regions, states and the European Union on ideas for rural community capacity-building. In this way, moves to exchange ideas, concepts and policy initiatives can be more realistically thought through, to the benefit of individual communities and indeed the local residents. Second homes will continue to be a vexed issue for most regions; what is important is a commitment on the part of policy-makers across Europe to exchange experiences, fostering new research to analyse the issue, but perhaps above all to incorporate the views of residents throughout policy development by initiating community-led discussions.

Bibliography

Action for Communities in Rural England (1988), *Who can afford to live in the countryside? Access to Housing Land*, Cirencester: ACRE

Aitchison, J. & Carter, H. (1985), *The Welsh Language 1961-1981: An Interpretative Atlas*, Cardiff: University of Wales Press

Aitchison, J. & Carter, H. (1986), 'Language areas and language change in Wales, 1961-1981', in Hume, I. & Pryce, W.T.R. (eds.), *The Welsh and Their Country*, Llandysul: Gomer Press

Albarre, G. (1977), The Impact of Second Homes: A. Second Homes and Conservation in Southern Belgium' in Coppock J.T., (ed.), *Second Homes: Curse or Blessing?*, Oxford: Pergamon Press

Aldskogius, H. (1967), 'Vacation House Settlement in the Siljan Region', *Geografiska Annaler* 49, 2, pp.69-95

Aldskogius, H. (1968), 'Studier i Siljansområdets fritidsbyggelse', *Geografiska Regionstudier*, 4, Sweden: Kulturgeografiska Institutionen vid Uppsala Universitet

Allen, C., Gallent, N. & Tewdwr-Jones, M. (1999), 'The limits of policy diffusion: second home experiences in Sweden and the UK', Environment & Planning C: Government and Policy 18, 2

Ambrose, P. (1974), *The Quiet Revolution*, London: Chatto & Windus

Archer, B. (1973), *The Impact of Domestic Tourism*, Bangor Occasional Papers in Economics 2, Bangor: University of Wales Press

Audit Commission (1992), *Developing Local Authority Housing Strategies*, London: HMSO

Balchin, P. (1981), *Housing Policy and Housing Needs*, London: Macmillan

Barbier, B. (1968), *Villes et Centres des Alpes du Sud* in Editions Ophrys, Gap

Barke, M. & France, L.A. (1988), 'Second Homes in the Balearic Islands', *Geography*, 73, 319, pp.143-145

Barke, M. (1991), 'The growth and changing pattern of second homes in Spain in the 1970s', *Scottish Geographical Magazine*, 107, 1, pp.12-21

Barlow, J. & Chambers, D. (1992), *Planning Agreements and Affordable Housing Provision*, University of Sussex, Brighton, Centre for Urban and Regional Research

Barlow, J., Cocks, R. & Parker, M. (1994), *Planning for Affordable Housing*, London, HMSO

Barr, J. (1967), 'A two home democracy?', *New Society*, 10, 7th September, 1967

Bennett, R.J. (1985), 'The impact on city finance of false registration in second homes: the case of the 1981 Austrian Census', *Tijdschrift Voor Economische en Sociale Geografie*, 76, 4, pp.298-309

Bennett, S. (1976), *Rural Housing in the Lake District*, Lancaster, Lancaster University

Bielckus, C.L., Rogers, A.W. & Wibberley, G.P. (1972), *Second Homes in England and Wales: A study of the distribution and use of rural properties taken over as second residences*, Wye College, London, School of Rural Economics and Related Studies

Bielckus, C.L. (1977), 'Second Homes in Scandinavia', in Coppock, J.T., (ed.), *Second Homes: Curse or Blessing?*, Oxford: Pergamon Press

Bishop, K. & Hooper, A. (1991), *Planning for Social Housing*, London: Association of District Councils

Bollom, C. (1978), *Attitudes and Second Homes in Rural Wales*, Cardiff: University of Wales Press

Bonneau, M. (1973), 'Résidences secondaires et tourisme en Maine-et-Loire', *Bulletin de la Societé Languedocienne de Géographie*, 7, pp.307-320

Bontron, J-C. (1989), 'Equipement et cadre de vie', Brun, A. (ed.), *Le Grand Atlas de la France Rurale*, Paris: de Monza

Bowen, E.G. & Carter, H. (1975), 'Some preliminary observations on the distribution of the Welsh language at the 1971 Census', *Geographical Journal*, 140, pp.43-142

Bramley, G. (1991), *Bridging the affordability gap in Wales: a report of research on housing access and affordability*, Cardiff: House Builders Federation and Council of Welsh Districts

Bramley, G. (1993), 'The enabling role for local authorities: a preliminary evaluation', in Malpass, P. & Means, R., (eds.), *Implementing Housing Policy*, Buckingham: Open University Press

Brier, M.A. (1970), *Les Résidences Secondaires*, Paris: Dunod Actualités

Brody, H. (1973), *Inishkillane: change and decline in the west of Ireland*, London: Allen Lane

Buchanan, R. (1985), 'Rural housing: romantic image in need of repair', *Radical Scotland*, 17, pp.22-23

Buller, H. & Hoggart, K. (1994a), *International Counterurbanisation: British Migrants in Rural France*, Aldershot: Avebury

Buller, H. & Hoggart, K. (1994b), 'The Social Integration of British Home Owners into French Rural Communities', *Journal of Rural Studies*, 10, 2, pp.197-210

Burby, R.J., Donnelly, T.G. & Weiss, S.F. (1972), 'Vacation home location: a model for simulating residential development of rural recreation areas', *Regional Studies*, 6, 4, pp.421-439

Champion, A. & Townsend, A.R. (1990), *Contemporary Britain: A Geographical Perspective*, London: Edward Arnold

Chartered Institute of Housing (1992), *Housing: the first priority*, Coventry, CIH

Clark, G. (1982), *Housing and Planning in the Countryside*, Chichester: Research Studies Press

Cloke, P. (1985), 'Wither rural studies?', *Journal of Rural Studies*, 1, 1, p.2

Cloke, P. (1996), 'Housing in the open countryside: windows on "irresponsible planning" in rural Wales', *Town Planning Review*, 67, 3, pp.291-308

Cloke, P. & Milbourne, P. (1992), 'Deprivation and lifestyles in rural Wales: rurality and the cultural dimension', *Journal of Rural Studies*, 8, pp.359-371

Cloke, P., Milbourne, P. & Thomas, C. (1994), *Lifestyles in Rural England*, Salisbury: Rural Development Commission

Cloke, P., Phillips, M. & Rankin, D. (1991), 'Middle-class housing choice: channels of entry into Gower, South Wales', in Champion, T. & Watkins, C., (eds.) *People in the Countryside: Studies of Social Change in Rural Britain*, London: Paul Chapman Publishing

Clout, H.D. (1969), 'Second homes in France', *Journal of the Town Planning Institute*, July, 1969

Clout, H.D. (1973), '350,000 second homes', *Geographical Magazine*, 45(10), p.750

Clout, H.D. (1977), 'Résidences Secondaires in France', in Coppock J.T., (ed.), *Second Homes: Curse or Blessing?*, Oxford: Pergamon Press

Coppock J.T. (1977), 'Second Homes in Perspective', in Coppock J.T., (ed.), *Second Homes: Curse or Blessing?*, Oxford: Pergamon Press

Coppock J.T. (1977), 'Issues and Conflicts', in Coppock J.T., (ed.), *Second Homes: Curse or Blessing?*, Oxford: Pergamon Press

Coppock, J.T. (1977), 'Social Implications of Second Homes in Mid- and North Wales', in Coppock J.T., (ed.), *Second Homes: Curse or Blessing?*, Oxford: Pergamon Press

Cribier, F. (1966), '300,000 résidences secondaires', *Urbanisme* 96-7, pp.97-101

Cribier, F. (1969), *La Grande Migration d'Été des Citadins en France*, Paris, Centre National de Recherche Scientifique

Crofts R.S. (1977), 'Self-catering Holiday Accommodation: The role of Substitution', in Coppock J.T., (ed.), *Second Homes: Curse or Blessing?*, Oxford: Pergamon Press

Crouchley, R. (1976), *Towards a Model of the Spatial Distribution of Second Homes in the UK*, University of Wales, Cardiff, Unpublished MSc Thesis

Cymdeithas Yr Iaith Gymraeg (1971), *Tai Haf* (Caerdydd, Cwmni Gwasg Rydd)

Cymdeithas Yr Iaith Gymraeg (1989), *Homes, Migration, Prices: Community Control of the Property Market*, Aberystwyth: The Welsh Language Society

Dartington Amenity Research Trust (1971), *The Gower Coast* Totnes: DART

Dartington Amenity Research Trust (1977), *Second Homes in Scotland: a report to Countryside Commission for Scotland, Scottish Tourist Board, Highlands and Island Development Board, Scottish Development Department*, Totnes: DART Publication No.22

Davies R.B. & O'Farrell P. (1981), *An Intra-Regional Locational Analysis of Second Home Ownership*, Cardiff: Department of Town Planning, University of Wales

Dawes, R.M. (1972), *Fundamentals of Attitude Measurement*, New York: John Wiley & Sons

Department of the Environment (1991), *Circular 7/91: Planning and Affordable Housing*, London: HMSO

Department of the Environment (1992), *Planning Policy Guidance Note 3: Housing*, London: HMSO

Department of the Environment/Welsh Office (1992), *Planning Policy Guidance Note 7: The Countryside and the Rural Economy*, London: HMSO

Department of the Environment (1996), *Circular 13/96: Planning for Affordable Housing*, London: HMSO

de Vane, R. (1975), *Second Home Ownership: a Case Study*, Bangor: Occasional Papers in Economics, No. 6, University of Wales Press

Dourlens, C. & Vidal-Naquet, P. (1978), *Résidences Secondaires, Tourisme Rurale et Enjeaux Locaux*, Aix en Provence: Centre d'Etude du Tourisme

Dower, M. (1965a), 'The fourth wave', *Architects Journal*, 20 January, 1965, pp.123-190

Dower, M. (1965b), *The Fourth Wave: The Challenge of Leisure*, London: Civic Trust

Dower, M. (1977), 'Planning Aspects of Second Homes', in Coppock, J.T., (ed.), *Second Homes: Curse or Blessing?*, Oxford: Pergamon Press

Downing, P. & Dower, M. (1973), *Second Homes in England and Wales*, London: Countryside Commission, HMSO

Drabble, M. (1990), *Safe as Houses: An Examination of Home Ownership and Mortgage Tax Relief*, London: Chatto & Windus

Dunn, M., Rawson, M. & Rogers, A. (1981), *Rural Housing: Competition and Choice*, London: George Allen & Unwin

Edwards, A.L. (1957), *Techniques of Attitude Scale Construction*, Appleton-Century-Crofts, Inc.

Emmett, I. (1964), *A North Wales Village*, London: Routledge and Kegan Paul

European Union (1993), *Tourism 1993 Annual Statistics,* Luxembourg: Office for the Official Publications of the European Communities

European Union (1996), *Social Portrait of Europe*, Luxembourg: Office for the Official Publications of the European Communities

Eurostat (1994), *Facts through figures: a statistical portrait of the European Union*, Luxembourg: Office for the Official Publications of the European Communities

Fraser, R. (1991), *Working Together in the 1990s: A Guide for Local Authorities and Housing Associations*, Coventry: Chartered Institute of Housing

Gallent, N. (1997a), 'Improvement grants, second homes and planning control in England and Wales', *Planning Practice and Research*, 12, 4, pp.401-10

Gallent, N. (1997b), 'Planning for affordable rural housing in England and Wales', *Housing Studies* 12, 1, pp.145-155

Gallent, N., Tewdwr-Jones, M. & Higgs, G. (1998), 'Planning for residential tourism in rural Wales', *Contemporary Wales*, 10

Gardavský, V. (1977), 'Second Homes in Czechoslovakia', in Coppock, J.T., (ed.), *Second Homes: Curse or Blessing?*, Oxford: Pergamon Press

Getimis, P. & Kafkalas, G. (1992), 'Local development and forms of regulation: fragmentation and hierarchy of spatial policies in Greece', *Geoforum*, 23, pp.73-83

Goodlad, R. (1993), *The Housing Authority as Enabler*, Coventry: Chartered Institute of Housing and Longman

Häbermas, J. (1987), *The Theory of Communicative Action: Vol 1, Reason and Rationalisation of Society*, translated by Thomas McCarthy, Cambridge: Polity Press

Häbermas, J. (1991), *The Theory of Communicative Action: Vol 2, Lifeworld and System: a critique of functionalist reason*, translated by Thomas McCarthy, Cambridge: Polity Press

Hansard (1985), *Housing surplus in England and Wales*, Written Answer, Col. 16, London: HMSO

Hall, J. (1973), 'Europe's seaside: landscape for leisure', *Built Environment*, 2, pp.173-175

Henshall, J.D. (1977), 'Second homes in the Caribbean', in Coppock, J.T., (ed.), *Second Homes: Curse or Blessing?*, Oxford: Pergamon Press

HM Government (1996), *Housing and Construction Statistics No. 63 (September Quarter 1995)*, London: HMSO

Hoggart, K. & Buller, H. (1995), 'British home owners and housing change in rural France', *Housing Studies*, 10, 2, pp.179-198

Hoggart, K., Buller, H., & Black, R. (1995), *Rural Europe: Identity and Change*, London: Arnold

Hughes, R.E. (1973), *The Planning Implications of Second Homes* Unpublished MSc thesis, Edinburgh: Department of Town and Country Planning, Edinburgh College of Art/Heriot-Watt University

Hutton, R.H. (1991), 'Local needs policy initiatives in rural areas: missing the target?', *Journal of Environment and Planning Law*, April 1991, pp.303-311

Instituto Nacional de Estadistica (1983), *Censo de viviendas, 1981, tomo IV: resultadas a nivel municipal* (Madrid), p.9

Ireland, M. (1987), 'Planning policy and holiday homes in Cornwall', in Bouquet, M. & Winter, M. (eds.), *Who From Their Labours Rest? Conflict and Practice in Rural Tourism*, Aldershot: Avebury

Jacobs, C.A.J. (1972), *Second Homes in Denbighshire*, Ruthin: County of Denbigh Tourism and Recreation Report No.3

James, C. & Williams, C. (1997), 'Language and planning in Scotland and Wales', in Macdonald, R. & Thomas, H. (eds.), *Nationality and Planning in Scotland and Wales*, Cardiff: University of Wales Press

Jenkin, N.M. (1985), *Towards a Policy Package for Second Homes in Anglesey*, Unpublished Diploma thesis, Cardiff: University of Wales Cardiff

Joseph Rowntree Foundation (1994), *Inquiry into Planning for Housing*, York: Joseph Rowntree Foundation

Jurdao Arrones, F. (1979), *Espana en Venta*, Madrid: Editorial Ayuso Madrid

Kemeny, J. (1992), *Housing and Social Theory*, London: Routledge

Kemeny, J. (1995), *From Public Housing to the Social Market: Rental policy strategies in comparative perspective*, London: Routledge

Larsson, G. (1969), *Undersökningar rörande fritidsbebyggelse* Institutionen för fastighetsteknik, sekt. Lantmäteri, Tekniska högskolan, Stockholm, Bulletin No.6, Part 4

Le Roux, P. (1968), 'Les residences secondaires de Français en Juin 1967', *Etudes et Conjoncture*, Supplement 5, 1968

Lewes, F. (1970), *The Holiday Industry of Devon and Cornwall*, London: HMSO

Ljungdahl, S. (1938), 'Sommarstockholm', in *Ymer*, 38

Llywelyn, E. (1976), *Adfer a'r Fro Gymraeg*, Pontypridd: Modern Cymreig Cyf.

Loughlin, M. (1984), *Local needs policies and development control strategies. An examination of the role of occupancy restrictions in development control*, SAUS Working Paper No.42, Bristol: University of Bristol

Martin, I. (1972), 'The second home dream', *New Society*, 18th May 1972

Martin, I. (1978), 'The impact of second homes: some comments on the experience in Wales', in Talbot, M. & Vickerman, R.W., (eds.), *Leisure Studies Association Conference Paper No. 8*, London

Mayle, P. (1989), *A Year in Provence*, London: Hamish Hamilton

Mayle, P. (1990), *Toujours Provence*, London: Hamish Hamilton

Messenger, J.C. (1969), *Inis Beag: Isle of Ireland*, New York: Holt, Rinehart & Winston

Mitchell, L.J. v The Secretary of State for the Environment and The Royal Borough of Kensington and Chelsea (1993), Court of Appeal Judgement CO-834-92

Morris, A.S. (1985), 'Tourism and planning in Spain with particular reference to the Costa Brava', in *Proceedings of the First Joint Conference of Hispanists in Polytechnics and other Colleges and the Iberian Social Studies Association*, Newton, M.T. (ed.), p.112, Newcastle upon Tyne Polytechnic

National Federation of Housing Associations (1988a), *Building Your Future; Self-Build Housing Initiatives for the unemployed*, London: NFHA

National Federation of Housing Associations (1988b), *Self-Build*, London: NFHA

National Federation of Housing Associations (1994), *Working households: affordable housing and economic needs*, London: NFHA

Newby, H. (1980a), *Green and Pleasant Land?*, Middlesex: Harmondsworth, Penguin

Newby, H. (1980b), 'A one-eyed look at the country', *New Society*, 14 August 1980

O'Connor, P. (1962), *Living in Croesor*, London: Hutchinson

Offices of Population Censuses & Surveys (1992), *1991 Census: Definitions Great Britain*, London: HMSO

Office of Population Censuses & Surveys (1993), *1991 Census: Housing and Availability of Cars*, London: HMSO

Official Journal of the European Communities (1995), 'Opinion on the proposal of a Council Directive [in OJ/95/c236/06] on the collection of statistical information in the field of tourism', *Official Journal 95/c236/20*

Oppenheim, A.N. (1966), *Questionnaire Design and Attitude Measurement*, London: Heinemann

Osgood, C.E., Tannenbaum, P.M. & Suci, G.J. (1958), *The Measurement of Meaning*, Illinois: University of Illinois Press

Ouren, T. (1969), 'Fritig og feriemilj', *Norwegian Geographical Studies*, 8

Pahl, R. (1966), 'The social objectives of village planning', *Official Architecture and Planning*, 29, pp.1146-1450

Pahl, R. (1970), *Whose City?*, London: Longman

Palatin, G. (1969), 'Le dévelopment des résidences citadines dans la région Grenobloise', *Revue de Géographie Alpine*, 57, pp.747-757

Pardoe, A.R. (1973), Personal communication about survey of second homes in the Aberystwyth Rural District to J.T. Coppock

Pardoe, A.R. (1974), '*Social Implications of Second Home Development in Mid-Wales*', Paper presented to the IBG Conference, Norwich

Pilkington, E. (1990), 'Burning Issue', *Roof*, 15, 2, pp.18-19

Psyhogios, T.E. (1980), *Second Homes in the Greater Area of Athens*, Unpublished MSc Thesis, Cardiff: University of Wales

Pyne, C.B. (1973), *Second Homes*, Caernarvonshire County Planning Department

Ragatz, R.L. (1970), 'Vacation homes in the North-Eastern United States: seasonality in population distribution', *Annals of the Association of American Geographers*, 60, pp.447-455

Ragatz, R.L. (1977), 'Vacation Homes in Rural Areas: Towards a Model for Predicting their Distribution and Occupancy Patterns', in Coppock, J.T., (ed.), *Second Homes: Curse or Blessing?*, Oxford: Pergamon Press

Reader's Digest (1970), *Survey of Europe Today*, Reader's Digest

Robbins, L. (1930), 'The elasticity of demand for income in terms of effort', *Economica*, June 1930, pp.123-129

Robertson, R.W. (1977), 'Second Home Decisions: The Australian Context', in Coppock, J.T., (ed.), *Second Homes: Curse or Blessing?*, Oxford: Pergamon Press

Rogers, A.W. (1971), 'Changing land-use patterns in the Dutch polders', *Journal of the Royal Town Planning Institute*, 57, 6, pp.274-277

Rogers, A.W. (1977), 'Second Homes in England and Wales: 'A Spatial View', in Coppock, J.T., (ed.), *Second Homes: Curse or Blessing?*, Oxford: Pergamon Press

Ross and Cromarty County Planning Department (1972), *Holiday Homes: Progress Report*, Dingwall

Rural Development Commission (1995), *Section 106 Agreements and private finance for rural housing schemes*, Salisbury: RDC

Salletmaier, C. (1993), 'The development and superimposition of tourism - Second homes and recreation within the rural fringe of an urban center', *Mitteilungen der Osterreichischen Geographischen Gesellschaft*, 135, pp.215-242

Saunders, P. (1984), 'Beyond housing classes: the sociological significance of private property rights in the means of consumption', *International Journal of Urban and Regional Research*, 8, pp.202-227

Savage, M. (1989), 'Spatial difference in modern Britain', in Hamnet, C., McDowell, L., & Sarre, P. (eds.), *Restructuring Britain: The Changing Social Structure*, London: SAGE

Shucksmith, M. (1981), *No Homes for Locals?*, Farnborough: Gower Publishing

Shucksmith, M. (1983), 'Second homes: a framework for policy', *Town Planning Review*, 54, 2, pp.174-193

Shucksmith, M. (1985), 'Public intervention in rural housing markets', *Planning Outlook*, 28, 2, pp.70-73

Shucksmith, M. (1990a), *Housebuilding in Britain's Countryside*, USA & Canada: Routledge

Shucksmith, M. (1990b), 'A theoretical perspective on rural housing: housing classes in rural Britain' *Sociologia·Ruralis*, 30, 2, pp.210- 229

South West Economic Planning Council (1975), *Survey of Second Homes in the South West*, London: HMSO

Soley, C. (1990), 'Seconds out', *Roof*, 15, 2 pp.38-39

Statens Offentliga Utredningar (1964), *Friluftsliver i Sverige*, Part I: Utgångsläde och utvecklingtstendenser

Tewdwr-Jones, M. (1995), 'Development control and the legitimacy of planning decisions', *Town Planning Review*, 66, 2, pp.163-81

Tewdwr-Jones, M. (1997), 'Land use planning in Wales: the conflict between state centrality and territorial nationalism', in Macdonald, R. & Thomas, H. (eds.), *Nationality and Planning in Scotland and Wales*, Cardiff: University of Wales Press

Tewdwr-Jones, M., Gallent, N., Fisk, M. & Essex, S. (1998), 'Developing corporate working approaches for the provision of affordable housing in Wales', *Regional Studies*, 32, 1, pp.85-91

The Guardian (1995), 'Housing: filling the empties', *The Guardian*, 26 July 1995

The Independent (1995), 'For sale: country pad for city slicker', *The Independent*, 17 August 1995

The Observer (1994), 'Welsh shed few tears as English beat a retreat', *The Observer*, 14 April 1994

Thompson, P. (1977), *An Investigation of Second Home Social Research Methodology*, Cardiff: University of Wales, Unpublished Diploma Thesis in Town Planning

Triandis, M.C. (1970), 'Attitudes and attitude change', Summers, G.F., (ed.), *Attitude Measurement*, New York: Rand McNally

Tuck, C.J. (1973), *Second Homes*, Merioneth Structure Plan, Subject Report No. 17, Merioneth County Council

Weatherley, R.D. (1982), 'Domestic tourism and second homes as motors of rural development in the Sierra Morena, Spain', *Iberian Studies*, 11, pp.40-46

Welsh Office (1982), *Response to the Memorandum submitted by Gwynedd County Council making proposals for action designed to limit the growth of second homes*, Cardiff, February 1982. The debate referred to was held on 11 December 1981

Welsh Office (1986), *Circular 30/86: Housing for Senior Management*, Cardiff: Welsh Office

Welsh Office (1988), *Circular 53/88: The Welsh Language: Development Plans and Planning Control*, Cardiff: Welsh Office

Welsh Office (1991), *Circular 31/91: Planning and Affordable Housing in Wales*, Cardiff: Welsh Office

Welsh Office (1992), *Planning Policy Guidance Note 3 (Wales): Housing*, Cardiff: Welsh Office

Welsh Office (1993), *Welsh Housing Statistics No.13*, Cardiff: Welsh Office

Welsh Office (1996a), *A Working Countryside for Wales: Rural White Paper*, Cardiff: Welsh Office

Welsh Office (1996b), *Planning Policy Guidance Wales*, Cardiff: Welsh Office

Whitehead, C. & Kleinman, M. (1991), *A review of housing needs assessment*, London: Housing Corporation

Wilcox, S. (1990), *The need for social rented housing in England in the 1990s*, Coventry: Chartered Institute of Housing

Williams, H. (1974), *Second Homes*, Cardiff: University of Wales, Unpublished Diploma Thesis

Williams, N.J. & Twine, F.E. (1994), 'Locals, incomers and second homes - the role of resold public sector dwellings in rural Scotland', *Scandinavian Housing and Planning Research*, 11, 4, pp.193-209

Wilson, A.G. (1974), *Urban and Regional Models in Geography and Planning*, London: Wiley & Sons

Wolfe, R.I. (1977), 'Summer Cottages in Ontario: Purpose-built for an Inessential Purpose' in Coppock, J.T., (ed.), *Second Homes: Curse or Blessing?* Oxford: Pergamon Press

For Product Safety Concerns and Information please contact our EU
representative GPSR@taylorandfrancis.com Taylor & Francis Verlag GmbH,
Kaufingerstraße 24, 80331 München, Germany

Printed and bound by CPI Group (UK) Ltd, Croydon, CR0 4YY
01/05/2025
01858546-0002